What if you could unscramble an egg?

What if you could unscramble an egg?

Robert Ehrlich

Rutgers University Press
New Brunswick, New Jersey

Library of Congress Cataloging-in-Publication Data

Ehrlich, Robert, 1938–
    What if you could unscramble an egg? / Robert Ehrlich.
        p.   cm.
    Includes bibliographical references and index.
    ISBN 0-8135-2254-4 (cloth)
    1. Science—Miscellanea.   2. Scientific recreations—Miscellanea.
3. Science fiction—Miscellanea.   I. Title.
    Q173.E47   1996
    500—dc20   95-18316
                                                                    CIP

British Cataloging-in-Publication information available

Published by Rutgers University Press, New Brunswick, New Jersey
Composition by Windfall Software using ZzTEX
Manufactured in the United States of America

Contents

What if grapefruit-size hail were common?
What if Earth were hollow?
What if Earth's atmosphere were much thinner?
What if Earth were always cloud-covered?
What if they dug a tunnel through Earth?
What if Earth's axis were tilted on its side?

Acknowledgments

I am greatful to the following people who gave me feedback on an earlier draft of this book: Anjuli Bamzai, Anthony Danby, Harvey Gould, Fern Levy, Darren Lewis, Robert Karlson, Karen Oates, Robert Sullivan, Charles Whitney, and especially my sister, Mildred Ehrlich.

**What if
you could
unscramble
an egg?**

_____

Chapter 1

Introduction

What if we had a conversation?

✧   A conversation about what?

✦   About anything under the sun.

✧   Could we narrow the subject a bit?

✦   Sure. Let's talk about all kinds of *what ifs* such as, What if people lived a thousand years? What if there were three sexes? What if you could go backward in time? Take a quick look at the headings throughout the book and you'll get the idea.

✧   What's the point of these questions?

✦   Sometimes, by looking at some of these bizarre possibilities, we can stretch our minds and better understand how our universe works.

✧   So why doesn't the author of this book just tell us how our universe works, instead of forcing us to have this stupid conversation?

✦   I rather enjoy our conversation. Maybe the author figures that by a free-ranging, somewhat flip conversation between two imaginary characters such as the two of us, he can bring up some of the same kinds of questions the reader might want to ask. The idea of explaining things through a dialogue has an honored tradition, going all the way back to Socrates of ancient Greece. Galileo, perhaps the first modern experimental scientist, also used a dialogue between two characters (Simplicio and Sagredo), to explain his theories of motion in his *Dialogues Concerning Two New Sciences.*

✧   Ah yes, I remember reading about my ancestor Simplicio—an engaging simple fellow, who was unafraid to ask whatever was on his mind. I gather Sagredo was your ancestor—a fellow who, like you, always had an answer for everything—although sometimes I wonder if he just made things up. So how shall we organize our free-ranging conversation? Let's organize the *what ifs* into a whole bunch of interrelated categories.

✦   OK, sounds good to me. We'll talk about some topics that are of general interest and some that are a little more technical. But we'll intersperse the two types; so some readers might enjoy skipping around. How about grouping our discussions under headings such as space, time, life, humans, the environment, disasters, physics. . . .

✧ Physics is boring. I don't want to talk about a whole bunch of complicated formulas, and ideas that have no relation to everyday life.

✦ Trust me, we'll talk about it in a way that even you find interesting.

✧ Are there *what ifs* we won't be talking about?

✦ Yes, lots of categories, like What if pigs could fly? After we've made a comment about the need to watch out for "presents" dropped by flying pigs, there doesn't seem to be a whole lot of interesting things to say.

✧ Well, we certainly don't want to bore our listeners with idle chitchat. What other kinds of topics won't we be talking about? I suppose we'll be avoiding the really hot potatoes?

✦ Not at all. There is no reason to avoid controversial subjects such as, What if there were three sexes? if we do it with intelligence and humor. But we will avoid ethical dilemmas, such as, What if you found out your spouse was cheating? or disasters such as, What if you got caught in an earthquake? We will instead be interested in bizarre possibilities such as, What if an earthquake occurred every day? or What if you were abducted by aliens?

✧ How did you know how I spent last week? But seriously, aren't some of these bizarre *what ifs* actually impossible—like the one about going back in time?

✦ We may find that some questions which sound impossible may in fact have some degree of possibility. The dividing line between the possible and the impossible may be blurrier than we care to admit. Anyway, even for the impossible ones, there is still some value in asking what the universe or society would be like if the *what if* were true. For example, by seeing why something is not possible, we can better understand why things are the way they are.

✧ So where should we begin?

✦ Let's begin where we all began—with sex!

Chapter 2

Sex

What if there were three sexes?

✧   Are there any species in nature with more than two sexes?

✦   Slime molds are said to have thirteen, but they still mate two at a time.

✧   That's good. Just the thought of a thirteen-party orgy gets confusing. Among humans, I guess you could say there actually are three or maybe even four sexes now.

✦   No, I'm talking about sex, not sexual orientation. But since you mentioned it, the diversity of human sexual behavior actually has been explained in terms of there being three sexes among the original version of humanity. The Greek philosopher Plato claimed that humans originally consisted of three types of joined pairs: male-male, male-female, and female-female. The god Zeus was said to have severed each pair so as to limit human powers. According to Plato, the various forms of sexual orientation can be understood as a desire to unite with one's missing half.

✧   It's not clear how bisexuals would fit into Plato's story. But if we return to the natural world, let's imagine a biology in which there were three sexes, in the sense of having three participants in the sex act. Can you imagine what those three roles might be?

✦   In addition to having one party supply the egg and the other supply the sperm, maybe there could be a need for some other ingredient to allow the egg to be fertilized.

✧   Sort of like a chaperon on a date?

✦   Actually, chaperons are supposed to stop the action from happening, so you'd better make that an "antichaperon," or what chemists would call a catalyst. Another possible third-sex role would be to handle the gestation, a job usually done by the female.

✧   But why would there be a need for nature to separate the egg-supplying and gestation roles into two sexes?

✦   Actually, there are some species that have precisely that division of labor. But nature hasn't found it necessary to have a third sex handle the gestation, because among all known creatures, gestation is handled by either the female or, more rarely, the male. Conceivably, the optimum conditions for producing sperm and eggs and for gestation might require very different body chemistry from that found in either males or females. In that case, a third sex would make biological sense.

✧   Sounds pretty far-fetched to me. Can you come up with any other reasons it might make sense to have three sexes?

◆   Yes—for error detection and correction. To verify that we have properly passed along the genetic instructions for making a new human being, it might make sense to have a third party involved in the sex act. To make an analogy, in order to prevent illegitimate transactions, banks require the insertion of two keys before a safe-deposit box can be opened.

✧   But what does that have to do with. . . . Oh, I see. But wouldn't that correspond to requiring two males to mate with one female, rather than having three different sexes?

◆   There would be only two sexes if all "keys" had the same basic shape, and had to fit into the same place in the "lock." If there were two different basic "key" shapes, and if each needed to be inserted in a different place, it would be reasonable to say there were three sexes.

✧   But why would nature require such a ridiculous thing?

◆   Like I said before—error detection. At present, nature sometimes makes mistakes in copying genes when half are taken from each parent and combined. We call these mistakes mutations. Even though the large majority of mutations are harmful, some are beneficial and some level of mutation is needed in order to allow evolution to work. But if we lived in a world where there were many more mistakes in copying genes—perhaps due to a much higher level of environmental radiation—it could make sense to have a third "player" in the sex act whose role was to detect and correct errors.

✧   What exactly are genes, anyway?

◆   The long DNA molecules found in each of our cells consist of sequences of coded building blocks. The smallest sequence associated with a particular inherited characteristic, anything from hair color to some component of intelligence, is called a gene.

✧   How might a third-sex player facilitate error detection when genes from each parent are copied?

◆   Among the genes to be passed along, all "normal" humans have certain sequences in common. The third sex could compare the new copies of these sequences with its own genes, and if it found a difference in one of the three versions, it would simply take the version that two of the three players shared.

This method wouldn't be foolproof, but it would catch most errors, which now are missed.

✧ What might society be like if there were three sexes?

✦ Whether or not all three sexes treated one another on an equal footing would depend on both cultural and biological factors. If there were roughly equal numbers of each sex, marriages might be three-person affairs. But, as in the present-day world, where polygamy is the norm in certain cultures, other arrangements might be possible. If, for example, there were very few members

of the "checker" sex, male-female couples might get together with checkers on special occasions.

✧   I wonder, if I explained all this to my two girlfriends. . . .

What if species could interbreed?

✧   I guess the possibility of interbreeding is how the ancient Greeks and Romans got their ideas for centaurs, satyrs, minotaurs, and other fantastic creatures. Why can't species interbreed, incidentally?

◆   The usual explanation is based on members of a given species having the same number of chromosomes, which are structures within the nucleus of each cell. In one sense, though, it's just a matter of definition—a species is the largest collection of similar plants or animals capable of interbreeding. In any case, the boundaries are sometimes blurry. Lions and tigers can interbreed, giving birth to ligers and tiglons—the former having a lion as father, and the latter a lioness as mother.

✧   Based on the Greek and Roman myths, I had pictured ligers and tiglons as having the front and back halves of lions. But it doesn't seem quite fair that we've forbidden species to interbreed simply based on a definition.

◆   You're right. Perhaps we should have phrased the original question as why the boundaries of a species aren't much wider than they are. Why, for example, aren't all mammals one species? I have to admit, though, that the idea of a mouse-elephant pairing does sound ludicrous—especially if it's a melephant, rather than an elemouse!

✧   True. But how come we have some species, such as dogs, whose members sometimes come in widely different sizes? A mating between a Saint Bernard and a dachshund seems almost as ludicrous as that between an elephant and a mouse. So if they can't interbreed, how come they're considered to be the same species?

◆   Anyone who has ever owned a female dog who went into heat can attest to the unlikely suitors that sometimes come around and make an attempt—a dachshund–Saint Bernard pairing might actually be possible—though artificial insemination would make it more feasible. In any case, it's at the genetic level, rather than the gross anatomical level, that members of a species are most alike, as they have a very large percentage of their genes in common. Even humans and chimpanzees, two closely related but distinct species, have 97 percent of the genes are in common. In fact, from a genetic perspective, chimps may be closer to humans than they are to gorillas.

✧   If humans and chimps share 97 percent of their genes, how come we and they (usually) look and act so differently?

✦ To an extraterrestrial, we and chimpanzees would probably look much more alike than most people would care to admit. But a good fraction of the overlap in our genetic makeup probably reflects similarities at the cellular level shared by all living things on Earth. In fact, humans have roughly 50 percent of their genes in common with yeast.

✧ But if humans are that close to some other species, such as chimps, how come we can't interbreed with them?

✦ Our genes supply the instructions for building our offspring—just as blueprints do for building a house. Imagine blending together two sets of blueprints for two different houses. There wouldn't have to be too much difference between the blueprints for the structure to be unsound, or the creature inviable.

✧ For the sake of argument, let's imagine that what we now think of as different species could interbreed, and that humans could mate with chimpanzees,

Harry introduces his better half to the guests.

horses, lions, dogs, and other animals. What would life be like under those circumstances?

✦   Conceivably, not too different from what now exists. Just because something is possible doesn't mean it will be widespread. Probably, very few people would be attracted to animals of the opposite sex. Also, the current taboo against bestiality might exist (possibly even more strongly) if animal-human matings could lead to offspring. It is barely conceivable, even now, that human-chimpanzee breeding is possible, but it has never been tried on moral grounds—though I suspect if it were possible, some adventurous soul would have tried it by now and made the results known.

✧   But let's take the hypothetical example a step further, and ask what might happen if human-animal interbreeding were not only possible but became widespread.

✦   In that case, new types of animals would evolve. For example, suppose the offspring of a human and a gorilla wound up with the best features of both parents (the strength of the gorilla and the intelligence of the human). In a precivilized society, such creatures might survive in greater numbers than their rivals. The same outcome might occur even if such "best-features" offspring resulted only from a fraction of all matings. The end result over many generations might be that "pure" humans and gorillas would die out and the mixed creature would take over.

✧   So if all different types of animals could interbreed, are you saying that, in the end, we'd be left with only one homogenized type of creature that was a blend of all of them?

✦   Almost certainly not, because the optimum conditions for survival vary greatly in different environments, so the genetic background that would allow one type of animal to thrive in one environment would make it poorly suited for others. In addition, even in a given environment, there are many ways to make a living. Biologists use the concept of a niche, which means that each organism has some specialized feature or behavior that allows it to take good advantage of certain resources, even if it may lack the strength, speed, or wits of its competitors.

What if men had babies?

✧   I suppose that male motherhood would quickly take care of the population explosion! But isn't it impossible? Wouldn't it just make men women, and women men?

✦   The division of labor between the two sexes can get a little confusing in some species. For example, among sea horses, after the females supply the eggs, the males fertilize and incubate them in a special pouch in their abdomen. The male sea horse gives birth to the young. The same idea was the basis of the movie *Junior*, except that the pregnant Arnold Schwarzeneggar didn't have a special pouch.

✧　Still, by definition, it's the female of the species that supplies the eggs, or in the case of mammals gives birth to live young—with the possible exception of some future experiment à la Schwarzeneggar.

✦　You're right of course about the defining characteristic of the female sex. But there often are other important secondary characteristics that differentiate the sexes. For example, among humans, men tend to be bigger, stronger, and more aggressive than women. Perhaps a better way to word the question is, What if the bigger and "nastier" sex were the one that had the babies?

✧　Now, let's not have any male bashing! Are there examples in nature where the female is the bigger and stronger sex?

✦　Sure, black widow spiders and praying mantises, to name just two. But among mammals that have an extensive period of child rearing and nurturing, the male is usually the bigger and stronger sex.

✧　Could it be that the demands of child rearing tend to make the female the more gentle member of the species?

✦　Probably so. Among humans, males are clearly, on the average, more aggressive and violent than females. Although it may be politically incorrect to say so, this male-female difference may have deep roots that go beyond the different way boys and girls are raised. There may well be innate, biologically rooted reasons why most Johnnies prefer war toys and most Janes prefer dolls—associated with the increased levels of the hormone testosterone in males. Attempts to make kids' behavior gender-neutral, however laudable, may have to overcome some natural tendencies in the opposite direction.

✧　In that case, how could it ever be that the stronger, more aggressive sex had the babies?

✦　Among some animals, such as penguins and many other birds, both parents share the work of feeding and nurturing the young. So the stronger, more aggressive sex could give birth, while the gentler sex actually raised the young.

✧　But why would it be advantageous to a species to have its stronger, more aggressive sex be the one either to have babies or to nurture the young?

✦　You haven't been around many two-year-olds, I take it. Actually, a number of scenarios would make this idea plausible. Suppose, for example, we lived in an environment where the traditional male role of hunter-gatherer required no great skill, perhaps because food were very plentiful. Also, suppose there were many other animals around that were continually trying to make a meal of human infants in the care of females. In such an environment, natural selection might favor the development of strong, aggressive females.

✧ Coming back to our own species, and the present-day environment, is there any way human females might wind up being the more aggressive sex?

✦ Sure. If, over the course of a number of generations, men should find strong, aggressive women sexy, and women should find sensitive, wimpy men sexy, these types of male-female pairings would become increasingly frequent. Assuming that aggression and wimpiness have a genetic component, then natural selection would tend to favor these types of men and women.

✧ Maybe there's hope for me after all.

What if males died right after mating?

✧ Perish the thought. Does such a thing ever happen in nature?

✦ The Australian marsupial mouse apparently works himself into such a frenzy after finding his mate that he dies right after losing his virginity.

✧ I guess one other possibility would be that some males are killed by females right after doing it—like the praying mantis or the black widow spider. I've never understood why males would bother mating knowing what the immediate outcome would be.

✦ Maybe word doesn't get around very well among spiders, particularly because the males who are killed don't get to warn anyone else. Also, even if spiders could actually think, those smarties that did avoid mating would not pass their genes along, and so the behavior that would be genetically rewarded by natural selection would be to mate regardless of the outcome.

✧ But why do the female black widows do it?

✦ As a matter of fact, the belief that black widows always kill their mates is not correct. Why do they do it some of the time? I don't know. Maybe the female black widow is not too happy with the male's performance.

✧ That's a pretty nasty way of showing her unhappiness.

✦ More seriously, though, it may simply be that the male is basically useless as far as the female is concerned after he "serves his purpose," and so she might as well have a good meal. Also, the female sometimes has her meal *before* the mating occurs, which may be her way of avoiding undesirables.

✧ But how would this strange behavior serve the interests of the species?

✦ You are assuming that nature always arranges things so that species over the long run adopt behaviors that serve their best interests. Evolution doesn't work that way. For example, humans seem to be on the verge of overpopulating the planet to the point where disaster may loom, yet nature has not rewarded low population growth—just the opposite. Evolution has no goal of any kind for a species. Rather, individual organisms whose genes best suit them to their environment simply survive in greater numbers.

✧ So you mean that nature rewards those genetically based behaviors that result in the greatest number of offspring?

✦ Almost. The thing that counts is the greatest number of offspring who survive to the age of reproduction. After that, any weaning out of "unfit" organisms has no impact on the relative numbers in future generations. One way of describing this process is through the idea of the "selfish gene," meaning that genes try to maximize their propagation at the expense of other genes—although not consciously of course.

✧ But what has all this got to do with black widow spiders? How would it be in nature's interest for the female to eat the male after mating?

✦ Well, I already gave you one reason: the female gets a good meal out of a useless male. More importantly, remember that only some males meet this fate. Presumably, the males that are killed are physically less capable than those that escape. The escapees are able to mate with other females, and on average they will have a greater number of offspring.

✧ But wouldn't the female's *not* killing her mate tend to lead, over the long run, to more offspring? Why shouldn't nature reward that more benign behavior?

✦ Maybe because the male spider doesn't mate many times during his life anyway, so that getting killed after the first or first few times doesn't result in much of a reduction, on average. It could be that this small reduction in matings and number of offspring is more than offset by the advantage of having stronger male spiders be the ones who get to pass their genes along.

✧ How about if human males met the same fate as many black widow spiders—what would society be like in that case?

✦ I imagine recreational sex wouldn't be too popular! Few males would want to waste what might be their only chance to pass their genes along to future generations. Also, child rearing would probably be handled by single females, barring the unlikely event that most males wanted to raise children fathered by others.

What if homosexuality were the norm?

✧ Being heterosexual, I guess I'd be a "sexual deviant" in that case. But is there an example of a society where homosexuality is in fact the norm?

✦ No, although homosexuality does appear to be present in virtually all societies, and in many societies it is not unusual for heterosexuals to have at least one homosexual experience.

✧ I take it you are saying that the division between strictly homosexual and heterosexual is not so straightforward.

✦ Yes. Homosexuality is probably not a yes-or-no proposition, but rather most people's sexual orientation probably lies somewhere on a continuum. In fact, in a few societies it is expected that all boys undergo a homosexual phase. Among some cultures, such as the Sambia in New Guinea, it is traditional that adolescent boys have regular homosexual relationships with older boys. But when they are adults, all are expected to get married (to women) and change their sexual orientation. Most Sambians manage the transition with no more confusion than teenagers anywhere else.

✧ Which is to say they are a mess. How about homosexuality in adulthood? Is that ever considered the norm anywhere?

✦ Among some Native American societies, such as the Mojave, and some cultures in South America and Polynesia, there is an accepted role for transvestites. Some are even kept as wives. Also, among the ancient Greeks, homosexual relationships between older and younger men were completely accepted. Spartan soldiers often fought side by side with their lovers.

✧ With no bad effects on combat effectiveness, I assume. But if homosexuality ever did become the norm, how would human society reproduce itself?

✦ In a predominantly homosexual society, sex could still sometimes lead to new children. Presumably, there would still be a heterosexual minority who could take care of carrying on the species.

✧ But if heterosexuals were the ones having the children, how could the genes favoring homosexuality ever become dominant in society?

✦ We don't know for sure that homosexuality is genetically based, although there seems to be some evidence in that direction. But even if that is the case, we could imagine that homosexuals (with some reluctance) would take part in the reproductive process, along with heterosexuals. Having relations with the opposite sex might be viewed by them as the supreme sacrifice for their country—much like serving in the armed forces in time of war. One could even imagine other possibilities.

✧ Like what, for example?

✦ In a conceivable future in which fertilized human embryos might be brought to term outside a woman's body, we could imagine the possibility of a complete separation between sexual practices and the creation of babies. Many contemporary homosexual couples do have a desire to raise children, even if they aren't interested in doing what it takes to create them.

✧ Most heterosexuals in today's society find the idea of homosexuals raising children somewhat repugnant, because of the possibilities of sexual abuse, and the thought of indoctrinating kids into an "unnatural" lifestyle.

✦ Sexual abuse probably occurs just as often with heterosexual parents, and "indoctrination" is just a disapproving way of describing the process of instilling values with which we disagree. We might imagine that in a predominantly homosexual society, irrational prejudice against the heterosexual lifestyle might be so strong that only homosexual couples would be legally allowed to raise children.

✧ I think I'd stay "in the closet" with my heterosexuality in such a society.

What if the human gestation period were ten years?

✧ I suspect most women would stop having babies, and the human race would become extinct!

✦ Of course, it might be "only" the last year or two of pregnancy during which a woman was very obviously pregnant.

✧ I'm sure women would be greatly cheered by that possibility. Are there any species of mammals that have such a long gestation period?

✦ There are none that have a gestation period of anything approaching ten years. Pregnancies vary in length among different species from just under two weeks for opossums to a maximum of about two years for the Indian elephant.

✧ But why would nature ever subject one of her creatures to a ten-year gestation period?

✦ The purpose of gestation is to allow the fetus to develop in a protected environment. One reason for a much longer gestation period would be if the process of development in utero were a lot slower than now.

✧ Are there any other conceivable reasons (pun intended)?

✦ Another possibility would be if the world were a more hostile environment. In that case, we might find that natural selection favored pregnancies of longer and longer duration, which would allow the baby to survive better on its own after birth.

✧ Of course, extremely long pregnancies might make it less likely that the mother survived until the birth! Are there any other ways that a ten-year gestation might make sense?

✦ Yes. A particularly interesting one, and perhaps the most realistic, would be the idea of a delayed pregnancy. In most species fertilization of the egg occurs shortly after mating, but there are some bats that mate in the fall, and the sperm stays dormant in the uterus during the winter months.

✧ So we might imagine a female getting pregnant and not having the fetus develop until some triggering event occurred, like the family car becoming available.

✦ If the human gestation period actually were ten years, it might lead to a very interesting possibility of people being unaware of the connection between having sex and having babies. If you think that is outlandish, consider that such an awareness was actually not uncommon in near-modern times, especially during the Victorian era.

✧ That seems like a pretty far-fetched idea. How could people fail to notice the connection when one always followed the other?

✦ That's just it: one would not always follow the other. Many matings would not lead to pregnancies, and if they did, the duration would not always be exactly ten years. In any case, why would anyone have a reason to connect a birth to what they were doing ten years before—even if they remembered? It is quite possible that people would think of giving birth as something that just happened to many women around a certain age.

✧ But surely some people would observe the mating of other mammals, with their much shorter gestation periods, and figure out the facts of life that way.

✦ Not necessarily. They might be curious as to why nature had its creatures perform that strange behavior we call sex, but its connection to reproduction could conceivably escape their notice. To take an analogy in the field of disease, it has only been in recent years that the link between smoking and lung cancer has been demonstrated. A very long gap between a cause and an effect can easily make the connection difficult to observe.

✧ Well, I guess that could be one more reason nature might give humans a ten-year gestation period: so women wouldn't realize what they were in for when they mated!

What if asexual reproduction of humans were possible?

✧ Reproduction without sex—perish the thought. But don't we do this now when a woman makes a "withdrawal" from a sperm bank?

✦ No; although artificial insemination can lead to pregnancy without the sex act, the newborn baby does have a mother and a father, even if the dad's identity may be unknown. By asexual reproduction, or cloning, we mean the development of an organism from a single cell through the process of cell division.

✧ Isn't cloning and the rest of genetic engineering tampering with nature?

✦ Nature creates a clone any time a pair of identical twins develop from a single fertilized egg. You may have done cloning yourself if you ever replanted cuttings from plants to create new plants. Humans are continually tampering with nature when we raise crops or animals that suit our purposes.

✧ But it's one thing to clone a plant and quite another to clone a person. Has this been done yet with higher life-forms such as mammals?

✦ Yes. Scientists now routinely clone mice, thereby producing genetically identical strains of mice having certain desired characteristics.

✧ Have they cloned any that look like Mickey? But why do they want to do this?

✦ By using genetically identical mice in laboratory experiments, scientists can eliminate one source of random variation in their results. Psychologists, for example, are fond of studying how quickly mice can learn to run a maze. Usually they want to use genetically identical mice, so that faster maze running can be attributed to better learning rather than to intrinsically brainy mice.

✧ How about people? Surely the much greater complexity of humans makes cloning a much more difficult problem.

✦ Despite the obvious anatomical differences between mice and people, they actually are quite similar at the genetic or cellular level. In fact, scientists have now shown human cloning to be possible—although they haven't actually carried out the process, due to ethical considerations.

✧ Well, I don't know how we can be sure it's possible to clone people until we actually do it. But what, exactly, are the ethical concerns about it?

✦ Most people probably would find it immoral to create a human life without a mother or father, just for the sake of a scientific experiment.

✧ Shades of Dr. Frankenstein and his monster, I guess. But suppose the scientist doing the cloning raised the clone as if it was his or her own child. What would be wrong with that?

✦ Cloning and raising a single child might be one thing. But the urge to reproduce is perhaps one of the strongest of instincts. If human cloning is allowed, a compulsive cloner might want to clone thousands or millions of copies of him or herself.

✧ Would the end result be thousands or millions of identical people?

✦ Not necessarily, since human attributes and behavior depend on some mixture of genetic inheritance and environment. If the clones were raised differently, they would be much like twins separated at birth and raised apart. Such cases have been extensively studied, and eerie similarities have been found between some separated twins.

✧ I guess some future Dr. Frankenstein might love to study a thousand clones raised in a thousand different environments. What other ethical concerns are there about human cloning?

✦ It might be great if we could clone a thousand Einsteins, but what if we cloned a thousand Hitlers? What if cloning were done for those who were willing to pay the price? One could imagine a billionaire paying to have many copies of him or herself cloned and raised.

✧ But suppose, despite all the ethical concerns about cloning, people did reproduce asexually. What would be the consequences?

✦ If nature had never "invented" sex, the likelihood is that organisms would show very little evolution over time, as explained in the next essay. In fact, it seems likely that only primitive organisms would exist on Earth in the absence of sexual reproduction. On the other hand, having evolved to our present "advanced" state, if people now were to reproduce asexually, the opposite concern suggests itself, namely, people deliberately choosing the direction in which humans should evolve. We might easily imagine efforts to design and clone a master race of people supposedly possessing superior attributes who might not have much use for the rest of us.

What if everyone looked alike?

✧ Life surely would be confusing if we couldn't tell who is who. But could this actually happen in the normal course of events?

✦ As long as humans continue to reproduce sexually, their offspring will acquire genes from both sets of parents and will not be identical to either one. Even if the original population of humans were extremely uniform genetically, over time mutations would tend to increase the diversity in the gene pool, and the result would be different-looking people.

✧ But as different races and ethnic groups intermarry and raise children, won't humanity tend to produce people that look more and more alike?

✦ Not at all. If different races should intermarry extensively, there might be far fewer people with the physical characteristics now associated with one particular race, but the diversity in the gene pool would be as extensive as before.

✧ So if in the future everyone looked alike, I suppose we are talking about some cloning experiment carried to the ultimate extreme, with the only permitted offspring being the clones of, say, "Our Great Leader."

✦ That seems to be the most likely possibility. If everyone were cloned from one person, we would all be as close as identical twins and would probably think and act very much alike, as well as look alike. It seems likely that society would be regimented to an incredible degree with little room for individuality.

✧ It sounds as if you're describing a colony of ants rather than a human society.

✦ That's right. In fact, the reason why ant colonies are so cohesive is precisely because all the worker ants are clones produced by a single queen. If everyone is everyone else's identical twin, each individual puts the welfare of the colony ahead of its own.

✧   Aside from regimentation and loss of individuality, what other consequences would there be if everyone were identical clones?

✦   One of the advantages of genetic diversity is that when environmental conditions change, a species can readily adapt to its changing environment. Genes that correspond to characteristics better suited to the new environment are more likely to get passed down to future generations. For example, moths whose coloration closely matches that of trees in their area are less likely to become a tasty meal for birds. In nineteenth-century England, when trees in some areas became increasingly soot-covered, it was the darker-colored moths who survived and outbred their lighter-colored brethren—even though initially nearly all the moths were light-colored. The lesson is that species having a completely homogeneous genetic background would be extremely vulnerable to environmental changes.

✧   But ants in a colony seem to have little difficulty adapting to their environment, despite their genetic uniformity. Why would the situation be any different for people?

✦   Of course, each ant colony is cloned from its own queen. Different colonies have somewhat different genetics, and these differences allow some ant colonies to adapt better to their environment than others. The first life-forms on Earth reproduced asexually. It seems likely that the reason nature invented sex in the first place was to give organisms a better ability to adapt to changing environments.

✧   And here I thought sex was invented so amoebas wouldn't have to stay home alone with their *Playboy* magazines.

What if everyone *still* looked alike?

✧   How could we keep track of who was who?

✦   People of very different ages could obviously be distinguished. But even for people the same age, there would inevitably be some small environmentally based differences, and matters of personal preference—like how long you wore your hair, or whether you used deodorant. Even identical twins can usually be told apart by people who know them well.

✧   But recognizing the differences between two people you know well is a lot easier than the problem of recognizing who is who if everyone looks alike. How could we deal with this?

✦   Four possibilities come to mind. One would be that we would introduce some method of labeling people. Perhaps everyone would have their name and social security number tattooed on their forehead.

✧   Let's forget about that one; it reminds me of something out of the Nazi era. What are some other possibilities?

✦   One is that we would become adept at picking up on the tiniest facial differences in identifying people. We often find that people of another race with whom we have not interacted much "all look alike." This inability to distinguish faces of another racial group may not be due to prejudice but simply a lack of familiarity with many small differences in a given type of face.

✧   It still seems that if people really all looked alike, the problem of telling them apart would be insurmountable. What's your third possibility?

✦   We would rely primarily on one of our other senses to tell people apart—especially the sense of smell, which for dogs is the primary means of identifying one another.

✧   I don't think I'd want to do the kind of sniffing dogs do to recognize my friends. Anyway, if everyone looked alike, they might smell alike also. What's the last possibility?

✦ We wouldn't bother to tell people apart. In our relationships with other people, we would treat everyone the same, without distinguishing friends or relatives from strangers.

✧ So I wouldn't have to explain to my wife what I was doing in bed with a "strange" woman, because neither my wife nor I would know who was who.

✦ Actually you couldn't have a wife, since that would imply that you did have a specific type of relationship with a specific member of the opposite sex.

✧ Is it really possible to imagine a world where everyone treated everyone else the same?

✦ It does seem far-fetched, especially when you consider that in the real world looking alike is no guarantee that people treat each other respectfully. Some of the bloodiest conflicts now and throughout history have been between groups of people who looked nearly the same but differed in their political or religious beliefs. Looking alike is no guarantee that people will get along. But there might be a special case if everyone actually looked identical.

✧ How come?

✦ The only plausible way for everyone to look alike would be if they had the same genetic makeup, and somehow were all close relatives. It is well known that most parents of most species will defend their offspring to the death. The importance of blood ties is further illustrated by the statistic that the incidence of murder of children under two years of age is twenty times lower for biological parents than stepparents. We can conclude that nature seems to have programmed its creatures to act altruistically toward other members of their species with whom they are closest genetically. You could say that it is our genes' way of "using us" to ensure their own maximal propagation. You are just your genes' temporary receptacle.

✧ Why do I suddenly feel like a laundry bin?

Chapter 3

Aliens

What if aliens landed who looked just like us?

✧ There have been quite a few science fiction stories with this theme.

✦ But actually, this is probably the least plausible appearance extraterrestrials would have.

✧ Why is that?

✦ Two reasons. First, because the conditions on a particular planet would greatly influence the way in which a creature would evolve. For example, a bipedal creature from a high-gravity planet would be expected to have stocky legs to support its weight, while on a low-gravity planet it would have spindly legs.

✧ So why is it that on Earth we have both chunky-legged and spindly-legged creatures, if these features are determined by the strength of gravity?

✦ Because a creature's anatomical proportions are based on both its height and the strength of gravity. Generally, very large creatures like elephants tend to be chunkier than small ones.

✧ Besides varying planetary conditions, why else would you expect aliens to look different from us?

✦ Because much of evolution depends on the accumulation of never-to-be-repeated historical accidents. You can think of evolution as a journey with a series of forks in the road, each depending on the previous forks. It is highly unlikely that even with the same environmental conditions, the course of evolution on Earth would be repeated elsewhere. The very different appearances of animals evolving in separated places (such as Australia and Africa) illustrate this.

✧ So is there any plausible way aliens might look just like us?

✦ Well, some believers might take the human-looking appearance of aliens as clear proof that humans were created in God's image.

✧ In the science fiction stories, the aliens who look exactly like people have disguised themselves so as to pass for people or clouded people's minds to believe the aliens look human, usually for the purpose of taking over the world. Although, I've never quite understood why aliens would travel halfway across the galaxy just to take over our measly planet.

✦   But maybe Earth is a very special place. After all, we are the only planet in the universe that gave birth to Einstein, Elvis, and *you*. If aliens who looked just like us landed, perhaps we are their descendants, and the accepted theory of humans descending from apelike ancestors is incorrect. Perhaps the alien-descendants of the original colonists would be coming back for a visit—possibly alerted to our technological maturity by their receipt of radio or TV signals from some of our finer programs.

✧   So you think people who claim the ancient Egyptians may have had extraterrestrial help in building the pyramids might be onto something? Or that ancient gigantic drawings on the ground visible only from the air were probably done by aliens?

✦   No, not really. We should not think that just because we have trouble figuring out how or why ancient people did things, we have to believe in ancient astronauts. I'm only saying that if we had an extraordinary event occur, the landing of aliens that looked very similar to people, one plausible explanation of that event would be that we both had a common ancestor.

✧   I'll bet we were a "lab experiment" started by some alien high school biology students, and they are coming back to see how we turned out.

What if aliens landed who were of vastly different size?

✧   You mean what if they were either much taller or much shorter than us?

✦   Right, although if they came from a high-gravity planet, they might be much more spread out horizontally than us, so height might not be the best measure of size.

✧   Is it possible that aliens could have microscopic size?

✦   It seems highly unlikely that *intelligent* aliens could be microscopic, because the development of a brain would require some minimum size. Insect-size intelligent aliens might be conceivable, in the case of very large insects.

✧   I remember once reading a science fiction story where the aliens radioed ahead that they would be landing at a particular place and time. A huge reception was arranged at the appointed time and place, including a brass band and even hot dog concession stands. Despite continued alien transmissions about their having just landed, the reception committee scanned the sky and the countryside in vain.

✦   What happened to them?

✧   In their final broadcast the aliens described their last sight, of extremely tall, highly acidic grass all around and a huge opening cavern engulfing them. It turned out that the alien ship, which was less than an eighth of an inch long, landed on a hot dog and was eaten by one of the onlookers.

◆   That story might be right on the mark. It seems quite conceivable that if aliens who were vastly smaller than us landed, we might not even know about it. Or if we did find out about them, it seems likely we might treat them as a curiosity to be put into a box and studied, rather than be treated as equals.

✧   But aliens coming from some distant star would undoubtedly be way ahead of us technologically, in order to be able to make the trip. Wouldn't their advanced technology permit them to dominate us, despite their tiny size? It would be as if we encountered some very stupid creatures hundreds of times our size. With today's advanced weaponry, we probably would be able to prevail.

✦   The tiny aliens might well prevail against us—just remember Gulliver in Lilliput! But it is unclear if they could get us to take them seriously as equals before hostilities began. Maybe their advanced technology would go arm in arm (or tentacle in tentacle?) with advanced "alien relations" skills that would permit a peaceful contact to be made.

✧   How about the opposite case of aliens who were hundreds of feet high landing?

✦   Here the possibility of a peaceful dialogue being established would seem even more remote. As far as the aliens were concerned, we might be seen as tiny, not very bright creatures akin to gerbils. But who knows? Maybe the advancement of a species technologically has little to do with its intellectual or spiritual advancement—as our own history might seem to indicate. Maybe, though highly advanced technologically, the aliens might be no more advanced than us intellectually or morally.

✧   In which case they may simply have us over for dinner. As dinner? But is it realistic to imagine that highly intelligent life could evolve to be vastly different from us in size? After all, doesn't intelligence require some minimum brain size and complexity?

✦   Scientists who attempt to rank the intelligence of various species usually assign less weight to brain size than to the ratio of brain mass to body mass. On this basis, it is conceivable that a mouse-size creature with a grape-size brain might have more raw brain capability than us, for example.

✧   If aliens ever do land, I just want to be sure that we and our brains remain raw.

What if aliens who wanted to speak only with dogs landed?

✧   This idea seems pretty far-fetched, although probably not any more far-fetched than the science fiction movies in which the aliens speak English. It sure would be a comedown to humans, though, if aliens landed and found dogs to be the species most worthy of their attention. Dogs do seem to have a pretty limited "vocabulary."

✦   Ah, but you're looking at dogs' vocabulary in too limited a context, perhaps.

For example, it could be that the sense of smell was the aliens' primary source of sensory input. They might then have much to "talk" about with dogs concerning the odoriferous world around them—a world of which we are largely ignorant. Or perhaps the aliens, like dogs, could hear sounds having frequencies above the range of human hearing.

✧ But surely the aliens would soon tire of talk about sensory matters and would want to pursue conversation on a more intellectual plane. How could they find dogs superior to humans intellectually?

✦ I have known individual dogs that seemed smarter than some humans, but it does seem inconceivable that dogs generally possess a (well-hidden) superhuman intelligence. Maybe the aliens would only be seeking out the species that was their intellectual equal.

✧ But that seems absurd. Aliens who could travel halfway across the cosmos to reach us would require a technology vastly ahead of ours. Is it conceivable that such a technologically advanced civilization might, by some measure, be intellectually on the level of dogs? How could that possibly be?

✦ Intelligence is not a one-dimensional quality but comes in a number of types. Think of the stereotype of a person who is intellectually gifted regarding scientific matters but totally lacking in common sense, or the idiot savant capable of astounding mathematical calculations who is otherwise retarded intellectually.

✧ Despite those examples, it still seems hard to imagine that an alien could have a highly advanced scientific knowledge and yet have a dog's level of general intelligence. Might there be some other possible explanation for aliens having a canine level of intelligence?

✦ The simplest possibility would be that the aliens who landed were not the ones who originally planned the journey. It seems quite likely that the original aliens might well have taken a number of other species with them on their long journey to Earth—either as pets or possibly to raise for food. Perhaps the original aliens were wiped out during the journey by some disease that left their pets unscathed. If their ship were under full automatic control, the pets would be the ones to reach Earth alive.

✧ That "pet theory" of yours seems a possible scenario, except for one problem. Dogs, like most other nonhuman species of animals, have little interest in interacting with ("talking to") other species, except perhaps to tell them, "Keep off my turf."

✦ Good point. But my theory would perhaps apply better if the aliens had a somewhat greater intelligence than dogs. Chimpanzees in captivity, for example, have been known to adopt dogs and cats as pets. If it were left up to the aliens, I imagine they would prefer "talking" to their intellectual equals, rather than their inferiors or superiors—wouldn't you?

✧ Could there be any other reasonable explanations why aliens might have a subhuman intelligence?

✦ One other intriguing possibility would be that the original highly advanced aliens embarked on a journey lasting many generations, during the course of which they slowly evolved into an intellectually inferior species.

✧ This idea seems ridiculous. On Earth the course of evolution has always led to creatures having improved capabilities, including intelligence. How could evolution progress "backward"?

✦ Evolution doesn't necessarily lead to a species having greater intelligence over time. Instead, it favors those capabilities that have important survival value. In many environments high intelligence does have such survival value, but other, nonintellectual attributes might be much more important in other environments. Even within the general category of intelligence, the more primitive components we share with animals have far more survival value than our "higher" forms of intelligence. In most societies, for example, craftiness and cunning are

more important survival tools than an ability to master calculus or appreciate Beethoven.

✧ In what way might the environment onboard an alien ship lead to the decline of intellectual abilities over many generations?

✦ Suppose, when the aliens designed their ship, they put everything under automatic computer control, and all the needs of the crew were taken care of. The crew would therefore be completely freed from everyday concerns. They could then spend their days either in purely intellectual pursuits or, alternatively, on the "holodeck," which would provide a full virtual-reality experience of a G-, R-, or X-rated variety. If the aliens' preferences were similar to those of humans, you can guess how many would be likely to engage in each type of experience.

✧ So over the course of many generations, the aliens might lose interest in intellectual matters and might have little understanding of the science and technology underlying the workings of their ship—a technology they would merely take for granted. I think maybe you are not thinking just of aliens in suggesting this scenario.

✦ Yes. Given the likely prospect of virtual-reality technology as an entertainment vehicle, it seems that we humans may become increasingly absorbed by entertainment and have less and less use for educational pursuits. In an era of computer-animated CD-ROMs, the activity of reading books may become a quaint eccentric pursuit, comparable to studying Latin. We may find that despite the much-touted "knowledge explosion," ordinary citizens in future generations will understand less and less about their world.

✧ Well, I hope they will be able to keep up their end of the conversation with Lassie.

What if you concluded you had been abducted by aliens?

✧ Do I detect from the phrasing of the question some skepticism that maybe people who think they have been abducted really have not been?

✦ Your skepticism detector is in good working order. I can think of at least four alternative possibilities to a genuine alien abduction, starting with the possibility that you had some kind of hallucination, perhaps while you were under the influence of some controlled substance.

✧ I don't take controlled substances, but I suppose someone could slip me something without my being aware of it. What are some other possibilities?

✦ If you had some friends who enjoyed practical jokes, you shouldn't rule out the possibility that someone was pulling a great joke on you—especially if you had previously made known your interest in UFOs.

✧   That doesn't seem terribly likely. It certainly wouldn't be easy for my friends to fabricate an alien ship that would fool me, and it wouldn't account for the large number of reported abductions. What are some other alternatives?

✦   You could have dreamed the whole thing. In fact, descriptions of alien abductions often describe the initial part as being like a dream. Many people have had dreams that they initially thought to be reality.

✧   Well, I'm not too impressed with your alternative explanations so far.

✦   Still another explanation would apply to those alien abductees who recall the event only while under hypnosis. The human mind is incredibly suggestible while under hypnosis, and in experiments psychologists find that all kinds of false memories can be created.

✧   What about the common elements reported in many alien abductions? Doesn't that show that we are dealing with a real occurrence rather than some kind of delusion?

✦   One of the common elements is the great interest the aliens seem to show in human sexual practices and anatomy. It is probably no accident that sex also is the subject of many of our fantasies. Some reports of sexual abuse elicited under hypnosis long after the fact may likewise be attributable to fantasies that therapists help to create through the same mechanism.

✧   But don't the aliens reported by abductees all supposedly look very similar? Surely that couldn't be a coincidence.

✦   Most of us have seen enough depictions of aliens in movies or books on alien abductions to know how they are supposed to look, so it is not surprising that fantasies about an abduction would now have fairly similar-looking aliens.

✧   And how do you explain the "missing time period" described by alien abductees? They pass out drunk, I suppose?

✦   I don't know what happens, but I have often found that while daydreaming during a long car ride, I find that a long period of time passes without my being aware of where the time went.

✧   I once read about a married couple who claimed they had been abducted and later reported they had seen a detailed star map while in the alien craft. It was found that their reported star map exactly matched the positions of stars in our vicinity as they would be seen from the aliens' home planet. How could the couple possibly have made up such a map on their own?

✦   The so-called map you are referring to actually bore very little resemblance to the star field in our neighborhood, and the close agreement with any actual star pattern is just a myth.

✧ So are you dismissing the possibility of real alien abductions?

✦ Not at all. It is just that when faced with such an extraordinary possibility, it seems reasonable to accept as a more likely explanation one of the other alternatives we have considered, until compelling nonanecdotal evidence appears. The most impressive evidence would be a physical object from an alien ship. For example, it is frequently reported that the aliens implant some kind of miniature device under the abductee's skin, presumably in order to track them later. Surely an analysis of these devices (which some people supposedly have had removed) should indicate whether they might have an extraterrestrial origin.

✧ You are not using extraterrestrial psychology! If the aliens wanted to conceal their presence here, they would certainly stick with a technology familiar to us.

What if we tried to communicate with extraterrestrials?

✧ Barring a UFO landing and a "take me to your leader" scenario, I assume we are talking about radio communication.

✦ That's right. Given the distances to the nearest stars, and the limitation imposed by the speed of light, it seems unlikely that we will be making any contacts with aliens other than through radio communications anytime soon—assuming, of course, that there is somebody out there with whom to communicate.

✧ Since radio signals travel at the speed of light and the nearest star is 4.2 light-years away, many years could elapse between messages. It would be a pretty strange conversation.

✦ That's right. To start we might just want to broadcast a message of friendship and announce our presence as an intelligent technological civilization. Of course, we may have done that already, since some of our radio and TV broadcasts already may have reached other civilizations.

✧ In which case, you could probably forget about the "intelligent" part of our announcement. But suppose we tried to put together a message specifically intended for extraterrestrials. What might we say?

✦ We wouldn't need to tell them where we were located, since they could deduce that from the direction from which the signal came. We might want to tell them something about ourselves, such as what we are made of, what we look like, and how tall we are.

✧ How could we be sure this kind of information could be meaningfully decoded by another civilization?

✦ All matter in the universe appears to be made of the same kinds of elements as found on Earth. Any element, such as oxygen, can be specified in terms of the number of protons and electrons an atom contains—in this case eight. Although the names for the elements obviously have no universal meaning, their atomic numbers do. A possible message would therefore need to set out first our symbols for the numbers 1, 2, 3, 4, . . . , and then indicate which atomic numbers represent elements of which we are made.

✧ Has this ever been attempted?

✦ Yes, as a matter of fact. In 1974, at the dedication of the radio telescope in Arecibo, Puerto Rico, a message was broadcast toward a cluster of stars in the constellation Hercules. It should arrive in about 26,000 years. If there is anyone alive on Earth 52,000 years from now, they might get a response. Of course, if there is anyone alive on Earth then, they might have as much idea of the content of the original message as we would have of a message sent by early humans 52,000 years ago—namely, zero.

✧ Let's get back to the content of the message, and our ability to design one that could be understood. What makes us think that alien beings would be expecting a message describing what atoms we are made of or how tall we are? We might want to describe how our brains work (if only we knew), how long we live, or our companion creatures on Earth.[1] Moreover, the things that are most important about us are of a nonphysical character, and they are more difficult to express.

✦ You are right, of course. Without having a clue as to what an alien would be interested in knowing about us, even the most carefully crafted message is likely to be indecipherable at the other end—unless by chance the aliens think somewhat along the lines that we do.

✧ In which case, we should be sure to mention that we don't taste very good! Is there some way we could get some "practice" in communicating with extraterrestrials?

✦ The best way would be to try to communicate with those creatures on Earth who have shown signs of high intelligence—the whales and dolphins. Some of the songs of the humpback whale last more than a half hour, and a few of them appear to be completely repeatable. In terms of their complexity—technically, their "information content"—these whale songs are far richer than this book! Humans have a long way to go before we even begin to figure out what these songs mean.

✧ Of course, it might help if we stopped killing the whales first!

---

1. Communicating our lifespan to aliens could be a bit tricky, since the length of time we call a year (based on one revolution of Earth around the Sun), would have no meaning to aliens. Any time or distance interval would probably be communicated most easily either in terms of atomic dimensions or in terms of the frequency and wavelength used to send the message.

What if life existed
in space?

✧   Of course, if we or some extraterrestrials should travel around in spaceships there would be life "existing in space," but I assume you mean life existing without any artificial surrounding environment.

✦   Right. We are really asking whether life as we know it, or some other form of life, could exist in space, off the surface of a planet.

✧   I recall at least one science fiction movie in which "seeds" of some alien species drifted across space to reach Earth, leading to their taking over their human hosts.

✦   Yeah, I enjoyed *The Invasion of the Body Snatchers* too, especially the original, but did you know that some scientists have suggested extraterrestrial seeds as a possible scenario for the start of life on Earth?

✧   I thought that life on Earth was supposed to have started from chemical reactions in some primeval "chicken soup" mixture of complex molecules.

✦   That's the usual hypothesis, but who really knows? Earth in its early history was a pretty inhospitable place. From the fossil record, we know that the earliest life-forms appeared around 3.5 billion years ago. Some scientists wonder if there was enough time between the existence of a stable Earth environment and the first appearance of life for life to have arisen spontaneously.

✧   But getting back to life in space, how about complex organisms, and not just seeds—how might life develop the ability to sustain itself in space?

✦   The two possibilities are that space-based life would have originated either in space or else on the surface of a planet. In either case, there would be big problems.

✧   Like what?

✦   If the life originated on a planet and adapted to conditions on the surface of the planet, it seems that it would be poorly suited to life in space, unless the planet had extremely low gravity, and no water and no atmosphere—conditions that approximate a space environment.

✧   But how could life survive in the absence of water and an atmosphere?

✦   Some organisms, such as brine shrimp, can exist in dormant form without water for an indefinite period, and an atmosphere is necessary only for an organism that breathes, so neither of those problems would be fatal. So there might not be any showstoppers to the formation of life on a very low-gravity planet.

✧   How about the formation of life in space?

✦   Without the gravity of a planet, matter in space is likely to exist in very dilute concentrations, such as a cloud. It is very difficult to see how a cloud of material could organize itself into complex structures.

✧   But didn't our galaxy, solar system, planet, and even ourselves initially form precisely from such an amorphous cloud?

✦   That's exactly what cosmologists believe. Based on computer simulations, complex systems have been shown to evolve from simple ones spontaneously. So we cannot rule out even something as far-fetched as a living space-born cloud, as in Fred Hoyle's science fiction story, "The Black Cloud."

What if life were abundant throughout the universe?

✧   Being a Trekkie, the idea seems very plausible to me. What is the prevailing view of scientists?

✦   Most scientists would agree with you, despite the lack of evidence so far, UFO reports aside. The most promising possibility for the existence of life in the solar system was thought to be Mars, but the Viking landing found no evidence for life there. Likewise, flybys of other planets show no telltale signs, such as water vapor in the atmosphere.

✧   If all the evidence so far is negative, how come scientists are optimistic about the chances of life being abundant throughout the universe?

✦   The reason is one of sheer numbers: our galaxy has around 400 billion stars, and it is one among hundreds of billions of galaxies in the universe.

✧   But maybe Earth and our solar system is a very special place.

✦   That's conceivable, but all indications are to the contrary. Our Sun appears to be a typical star, and given what we know about the way stars are formed, it seems likely that many other stars have planets. In fact, we have even detected several planets around a special kind of star, a pulsar, from which we can detect radio pulses arriving at a regular rate. Sufficiently massive planets around such a star can reveal their presence by causing the time between the pulses to vary enough for us to detect them.

✧   Even if there are many stars with planets, how do we know life could evolve on them?

✦   Based on fossil evidence, life seems to have evolved on Earth as long as 3.5 billion years ago—a "mere" billion years after Earth's formation. Putting aside the possibility of an extraterrestrial origin of life, that time frame is about as early as life could have first appeared without being destroyed by the frequent large asteroid impacts during Earth's early history. Since life appeared so soon on Earth, and since the appearance of life does not seem to require any departure from the laws of physics and chemistry, many scientists think it should emerge anywhere that environmental conditions allow it.

✧ What might those conditions be?

✦ Stars like our Sun—the so-called yellow dwarfs—would appear to be the most promising places to find habitable planets. Those stars which have a much smaller or larger mass than our Sun, and hence a much lower or higher temperature, are much less promising places. For cooler stars, a planet would have to be in a much smaller orbit to get enough warmth for liquid water to exist, and strong tidal forces might cause catastrophic environmental conditions. On the other hand, hotter stars tend to burn at such a furious rate that they live very short lives—possibly too short for life to evolve. Aside from the temperature of the Sun, there are presumably lots of other factors influencing the probability of life evolving from complex nonliving matter that we can only guess at.

✧ But how could complex living systems evolve spontaneously from simpler nonliving systems? Don't things always go in the opposite direction (from order to disorder), if left to themselves?

✦ The idea of systems being required to increase their degree of disorder, or entropy, over time holds only for closed systems— that is, systems that are isolated from their environment. Among open systems, which take energy from their environment, there are many examples of order spontaneously emerging from disorder, or complex systems emerging from simpler ones. Although the detailed chemistry as to how life originated is not yet understood, many theories of how it might have happened seem plausible.

✧ How about life that is both intelligent and at a level of technology at least equal to us, which we can call Intelligent Technological Life (ITL); how abundant is ITL likely to be?

✦ Difficult to say. We really have no way of knowing whether the 3.5 billion years it took ITL to evolve on Earth is typical or not. Most importantly, we have no way of knowing how long the average lifetime of ITL is. It is quite conceivable, for example, that on average ITL tends to destroy its civilization after only a century—which is roughly the amount of time since we have become an advanced technological society.

✧ Why do you consider our technology to be roughly a century old?

✦ That is how long we have known about radio—which is very likely how another ITL would try to communicate with us—assuming they wanted to. We have, incidentally, been listening for such signals and not found any yet, but perhaps we don't know exactly what to listen for.

✧ How about UFOs? If life is abundant throughout the universe, isn't it likely that we are being visited regularly?

✦ Not likely. Despite the "warp drive" of science fiction, the speed of light does seem to be a universal speed limit. Given typical interstellar distances, it would probably take anywhere from many years to many centuries for the nearest ITL to make the trip—depending on your assumptions regarding the abundance of ITL. Also, even if only a fraction of the UFO sightings represent real extraterrestrial visitations, it is difficult to understand why such a primitive life-form as us would attract such a high degree of interest.

✦ Speak for yourself.

Chapter 4

Life

What if all living things were part of one giant organism?

✧   I have heard of the Gaia hypothesis—James Lovelock's idea that Earth itself and all its creatures can be considered a living organism. But how could this possibly be? Different creatures are physically separate, each living its own life.

✦   The same could be said for the individual cells of your body. These cells are individual living entities. It seems hard to imagine that your cells could have any awareness that collectively they make up a higher being—you.

✧   Suppose we pretend that the cells of your body do have conscious awareness, and could observe the events going on inside your body. Could they figure out that they were part of a higher life-form?

✦   Probably they would be in much the same situation we are. They might have indications they were part of a higher life-form, but they would not be able to point to anything conclusive. Let's look at what intelligent cells would find: a lot of related functions performed by different types of cells (a division of labor), and extensive networks of communication between widely separated cells.

✧   What else might be an indication to the cells that they were all part of a higher being?

✦   They could observe that although cells were continually dying and other cells continually being created, certain stable structures continued to exist (our organs), which seemed to perform specific functions that somehow reinforced one another. If the cells could monitor the body's environment, they might find that they often acted in concert to maintain stable environmental conditions, without any conscious desire to do so. In much the same way, living creatures on our planet (with the possible exception of humankind) generally maintain a stable environment without consciously cooperating.

✧   Yes, but individual living creatures often compete with one another rather than cooperate to achieve some higher goal, as the cells of our body usually do.

✦   But often the net result of that competition serves to promote the welfare of the species, as when males compete for a mate. In this case, when fitter males get to do the mating, superior genes are passed to future generations. So competition can promote a higher goal just as well as cooperation.

✧ What about warfare between humans, and between some animals? How can this be reconciled with the idea of all life being part of some higher being?

✦ Warfare between individuals in no way conflicts with the idea that they are part of a higher being. For example, warfare goes on inside our bodies continuously, as alien organisms continually enter the body.

✧ Aside from the analogy of the individual cells of our bodies, what other example can you give for a collection of individuals forming a higher life-form?

✦ Think of the ants or bees making up a colony. Although the behavior of individual ants is very simple, that of the colony taken as a whole is amazingly complex—just like the collection of living cells making up you. Moreover, the behavior of individuals is so "altruistic" that it is meaningful to think of the group as a single living organism, which myrmecologists (ant scientists) actually do in many contexts.

✧ If it were the case that all of life constituted a single living organism, what would be the place of individual humans, who often like to think of themselves as the highest form of life?

✦ Humans might be the brain cells or neurons of a higher life-form. The neurons of the brain have an extensive web of interconnections; each one can "talk" to many other remote neurons. With the advent of the coming information "superhighway," we may be approaching the point where everybody can talk to everyone else. At that point we might think of ourselves as being in a position similar to the individual neurons in some higher being's brain.

✧ So are you suggesting that the higher life-form of which all life is a part is still in the process of being formed?

✦ Quite possibly. The process of establishing such connections between neurons in our own brains is exactly what happens during the gestation period of a human child. The universe may be on the threshold of giving birth to some higher life-form.

What if scientists create life?

✧ Do you mean mixing a bunch of chemicals together and having something crawl out of the test tube?

✦ That's hardly likely, but perhaps we could imagine creating something much more primitive—a virus, for example, which is sort of halfway between living and nonliving.

✧ That's ridiculous. Something is either alive or it's not alive.

✦   Well, viruses, when they are outside a living cell, appear no more alive than many other complex molecules that scientists are now capable of synthesizing, but once inside a cell they "hijack" the cell's machinery and use it to make copies of themselves.

✧   All we need is for some fool scientist to create a new virus. How far have they actually gotten in trying to create life?

✦   Scientists as early as the 1950s showed that all the complex molecules on which life is based could be formed by natural processes. It was found that these complex molecules could be synthesized in an atmosphere similar to what probably existed on the young Earth, simply by the addition of energy in the form of electrical discharges, heat, or other forms.

✧   But how could complex living structures like even the simplest bacteria form from simpler nonliving structures? I thought the theory of "spontaneous generation" of life was long ago discredited.

✦   Actually, most scientists now believe that given a sufficient input of energy, and the proper environment and raw materials, it is virtually inevitable that a collection of complex molecules will assemble themselves into higher levels of structure—and possibly a primitive form of life.

✧   But even granted that a primitive form of life evolved at one place on Earth, how could it spread over the entire planet?

✦   There are many examples of chemicals that, in a favorable environment, produce substances that serve as catalysts for their own creation. Such "autocatalytic" reactions mean that once a little bit of something is produced, you don't have to wait long before you have quite a lot of it.

✧   Well, clearly scientists must be missing some "vital" ingredient, because they have not produced life in a test tube so far.

✦   The one missing ingredient is probably time. Nature had upward of a billion years or so to produce life on Earth, which leaves plenty of time for false starts.

✧   I guess it also had some pretty large "test tubes" to work in, assuming life first developed in the oceans. Can you come up with any other prospects for the artificial creation of life?

✦   Some people say that at the rate at which computers are improving, we may at some point be creating entities that are functionally alive, even if they are based on a totally different chemistry. It may not happen in the near future, but it is not difficult to imagine computer-controlled robots that would have many of our abilities.

✧   Surely you aren't suggesting that a robot with a computer-brain could be considered alive. Computers don't "understand" the words and symbols they input and output.

✦ In the future we may have supercomputers that can duplicate many, if not all, human mental tasks. It may be impossible to determine whether or not such supercomputers "understand" any less than we do. If that seems implausible, consider that we normally learn whether other people understand us through conversation. Computers have already been programmed to "converse" with people in a way that sometimes makes it difficult to tell whether a computer or a live person is present at the other end of the conversation. If such computer mimicry of normal conversation were to become perfect, on what basis could we know the computer doesn't understand us?

✧ I suppose that much of what passes for understanding in human affairs could be based on mimicry too. But aren't all living organisms based on the same carbon chemistry? The material composition of computers would seem to put them in an entirely different nonliving category.

✦ That's true as far as life on Earth is concerned. But why should we use carbon chemistry as an essential defining feature of life? The key defining characteristic of life may well be the way matter is organized, rather than the particular composition of the matter, be it carbon or, in the case of today's computers, silicon.

Research into the creation of life continues in its sexist tradition.

✧   Well I certainly would not want to be accused of being a "carbon-chauvinist!" If we do create life—either of the chemical or computer variety—will we have shown that God doesn't exist?

✦   Not at all. We could just as easily conclude that it was part of God's plan for His or Her creatures to create new life.

What if a life form between plants and animals existed?

✧   I'm not sure I know what it means to be "between" plants and animals. Maybe we should define our terms.

✦   Plants get their energy directly from sunlight through photosynthesis, and in the process they take in carbon dioxide and expire oxygen. Animals get their energy from eating plants or other animals, and they take in oxygen and expire carbon dioxide. Unlike plants, most animals move around and have specialized sense organs that allow them to react quickly to their environment.

✧   Trying to think what a life-form between plants and animals might be like reminds me of that science fiction musical *Little Shop of Horrors*. I take it we're looking for animals that have some of the properties of plants, and vice versa. How about the Venus's flytrap?

✦   That's a good one. The Venus's flytrap, which is native to a small area of North Carolina, is the most dramatic example of insectivorous plants. They can trap and digest insects and other small animals, which supply the plants with nutrients they can't get from the poor soil in which they tend to grow. The Venus's flytrap is a plant, however, since it does photosynthesize and take in carbon dioxide.

✧   What other categories of "intermediate" life forms are known?

✦   One category would be animals that are almost stationary and lack the kind of sense organs usually possessed by animals. Catatonics aside, such relatively stationary creatures could not exist on land, because unlike mobile animals, they would be unable to get lunch when they got hungry. But nearly stationary animals like clams, oysters, tube worms, and coral reefs can make a living on the seafloor, where a steady stream of nutrients drifts downward.

✧   Are there other life-forms with characteristics intermediate between plants and animals?

✦   One other would be the fungi, which many biologists consider a category separate from plants. Fungi do not get their energy from photosynthesis, but like animals they get it secondarily from plants, dead or living, on which they feed.

✧ Are there any combinations of plant and animal characteristics that are never found in nature?

✦ One would be a plant that wanders around like an animal but, being a plant, gets its energy from photosynthesis.

✧ With all the other combinations, how come we never see this particular one in nature?

✦ The ability animals have to move around requires much more energy than plants can generate through photosynthesis. When animals eat plants, they take in much more energy than plants do in the same time, because the animals essentially reap all the energy the plants have stored through photosynthesis over a period of time.

✧ So I guess a hypothetical plant that wandered around would not be able to take in enough energy through photosynthesis to allow it to keep moving. Would there be any other reasons why so few life-forms on land have characteristics intermediate between animals and plants?

✦ Plants and animals have coevolved on Earth to provide one another with everything they need. In this symbiotic relationship, plants provide animals with oxygen, food, shelter, and even medicines. In return, animals provide plants with carbon dioxide, fertilizer, and pollination. The relationship between the two life-forms has been a stable one that has allowed life to flourish.

✧ In that case, I'm glad there aren't any "planimals," or "animants."

What if dinosaurs had not become extinct?

✧ Didn't the fact that the dinosaurs became extinct mean that they were ill-suited to their environment?

✦ That's what we imply when we call somebody a "dinosaur," but actually the dinosaurs were very successful animals, having existed on Earth for 165 million years. At the time dinosaurs became extinct 65 million years ago, possibly as a result of an asteroid impact, so did about three-fourths of all species on Earth.[1] Larger animal species were hit especially hard. Of course, not all dinosaurs were extremely large, but even the smallest ones were larger than most mammals then alive.

---

1. The notion that the extinction of the dinosaurs was caused by an asteroid impact was suggested by Luis and Walter Alvarez. Their evidence consisted of a thin layer of the element iridium found in sedimentary rocks around the world. Iridium is rare on Earth but not in meteorites, and the depth of its layer placed the deposition at a time of 65 million years ago.

✧   Why should larger animals be disproportionately affected by an environmental catastrophe?

◆   It probably has to do with the fact that big fierce animals tend to be rare. Big animals that feed on smaller ones cannot be too numerous in a given area or they will starve. If a cataclysm occurs, such as the severe climate change believed to be responsible for the dinosaurs' demise, big animals tend to be most vulnerable to becoming extinct because they are relatively rare: a large drop in their population could easily leave too few to reproduce.

✧   Aren't the dinosaurs thought to have been rather stupid animals, not capable of posing a real threat to human ancestors, even if they had not become extinct before we came on the scene?

◆   The competition for survival between different species of animals is somewhat similar to a war. As in a war, victory can depend on a combination of fighting ability, offensive and defensive weaponry, intelligence, numbers, and luck. If large dinosaurs posed a threat much greater than other fierce predators that lived at the time of humans' emergence, it is conceivable they could have prevailed against human ancestors, even if dinosaurs are assumed to have had low intelligence.

✧   You don't really believe that dinosaurs could have prevented the emergence of humans, do you?

◆   Who knows? But as in any war, it is difficult to be sure what the outcome would have been if the "enemy" had gotten lucky at certain points. If you believe that man was preordained to have dominion over the creatures of the Earth, then the outcome could not be in doubt. But if you look at it as a competition for survival, then "victory" would go to the most successful predator.

✧   You make it sound as if humankind came out on top because we were the most successful killers.

◆   That's exactly right. Anthropologists have found that many apelike creatures existed during the period leading up to the emergence of humans. A very large number of these showed evidence of having been killed by primitive weapons. So we may have descended from the species that was most successful in killing off the competition.

✧   But getting back to the dinosaurs, could they have posed a serious threat to early humans, given their low intelligence?

◆   It is true that, based on their brain size, most dinosaurs were probably quite stupid, but at least one type, the small ostrich-like dinosaurs known as Saurornithoides, had a ratio of brain size to body weight that was comparable to mammals. So it is quite possible that the intelligence of dinosaurs would have evolved considerably further over time.

✧   Why would their intelligence be expected to improve over time?

✦   Some people think that early humans' brain capacity took a big leap forward at the time they began to walk upright, freeing their hands to manipulate their environment and build tools. In other words, rather than saying that a big brain allowed us the capacity to make tools, you could turn it around and say that hands that had a toolmaking ability made it more advantageous to have a big brain.

✧   So in other words, given that many dinosaurs also walked upright with hands free, we should expect that an increase in their brain size would have occurred naturally.

✦   Exactly. We could even speculate that were it not for the dinosaurs' extinction, their descendants would have won the "arms race" in intelligence against the mammals, and those dinosaur descendants would be the ones writing books and speculating about how things might have turned out differently.

What if we could
bring the dinosaurs
back to life?

✧   Do any scientists think that the scenario in *Jurassic Park* could actually happen?

✦   Some think that it might be possible one day. In the movie, dinosaur DNA is recovered from a mosquito that had presumably bitten a dinosaur and gotten stuck in some tree sap, which over the course of millennia became fossilized in a form we call amber.

✧   Could this first step of the dinosaur-building process actually be done today?

✦   With considerable effort, scientists have already extracted DNA from a weevil preserved in 130-million-year-old amber. Even if a weevil might have bitten a (baby?) dinosaur, it would be tough to disentangle the weevil DNA from that belonging to any dinosaur it might have bitten. Moreover, if we did accomplish the separation, we wouldn't have any idea which dinosaur the DNA belonged to. Also, remember that, contrary to popular image, most dinosaur species were fairly small creatures, rather than the hulking monsters we associate with the word.

✧   Isn't there some way we could get DNA for a particular dinosaur species—from its bones, perhaps?

✦   DNA has been found in ancient bones, but only very chopped up, with many missing pieces. Scientists now do have ways of filling in missing segments with any desired sequence, but we have no idea what a dinosaur DNA sequence is supposed to look like. One approach to filling in any missing pieces would be to splice the dinosaur DNA onto that of the dinosaurs' closest living relatives.

✧   Would that be some of the reptiles?

✦   No, actually, it would be the birds. An ostrich might be the best choice, given its anatomy and size.

✧   It sounds like we might wind up with the blueprint for making something between a dinosaur and a bird, rather than a true dinosaur. But let's suppose we could get some dinosaur DNA. How could we use it to create the animal?

✦   That's the really tricky part. Let's say we wanted to find some way to have an ostrich lay a dinosaur egg. In principle, what you would want to do is to take the sperm and egg cells from a pair of ostriches and replace the ostrich DNA in the nuclei of those cells with dinosaur DNA. Unfortunately, we have no idea how to do this without killing the cells in the process.

✧   Suppose we could somehow solve that problem, and the ostrich actually did lay a dinosaur egg (which sounds pretty painful just to think about). Could we raise the dinosaur without a mother?

✦   That's the easy step. We do it now for nearly extinct species, such as the condor. There are so few of these birds left that we don't want to let nature take its course.

✧   But raising an animal we have never observed in the wild poses all sorts of special problems, like knowing what they like to eat.

✦   Some scientists have made a study of fossilized dinosaur poop. Of course, it is not always easy when you find some to tell for sure who produced it. But based on these studies, we can learn what the dinosaurs ate, and find some modern equivalents.

✧   It sounds like the bottom line is that we don't know how to bring the dinosaurs back now, but the idea might not be out of the question someday. If we ever did have the capability, should we do it?

✦   If we did it just for the sake of creating, in effect, a fantastic amusement park or zoo, why not do it using dinosaur robots, which would probably be a lot easier and cheaper. Of course, they wouldn't be real dinosaurs, but then again, anything we created by splicing together dinosaur genes with those of ostriches might not be "real" dinosaurs either.

Chapter 5

Humans

What if a pill could give you knowledge?

✧ If scientists invented a knowledge pill, presumably, kids wouldn't have to go to school, and mothers everywhere would rise up in arms to keep the pill off the market! But, seriously, there isn't any prospect of such a development, is there?

✦ We have no indication that such a thing could be done, but based on what we know about the chemical basis of memory, we can't say it is impossible. In one study, for example, flatworms were trained to turn toward a light. When the trained worms were ground up and fed to some new worms, the new worms learned faster than the worms who didn't receive this diet.

✧ I've always had a hard time learning physics. Perhaps there might be an effective use for some of my old professors after all.

✦ Of course, we haven't a clue how abstract human thoughts are stored in the brain, so what might work for flatworms might have little bearing on humans. On the other hand, we do know that many drugs affect brain functioning, so it is not impossible that memories could be created or altered by a pill.

✧ I think I'd stay clear of such mind-enhancing drugs if they worked in any way like the drugs to enhance physical abilities.

✦ You're right about the harmful side effects of steroids. In fact, most drugs have side effects of one kind or another, some of which are discovered only after long use. With mind- or memory-enhancing drugs, the potential side effects could cover a much wider range, encompassing all aspects of mental functioning. You might take a pill to learn nuclear physics, only to discover that you developed the personality of a nerd, or a craving for chocolate-covered ants!

✧ Even apart from side effects, could the knowledge gained from a pill really compare to that obtained through thinking?

✦ At its deeper level, knowledge of any subject is far more than a collection of facts; it is a web of interconnections between various abstract ideas. To know whether or not we could acquire such knowledge through drugs, we would need to increase greatly our understanding about how our brains work. Until such understanding of the brain is achieved, perhaps the most likely possibility might be mind-enhancing pills that could improve our ability to learn new information.

✧   OK, but let's say in the distant future we could actually take a pill to gain new knowledge without any effort on our part. What would be the likely impact on society?

✦   That would greatly depend on who designed the pills, what kind of knowledge they included in the pills, who controlled their distribution, and how much the pills cost. One could imagine all kinds of nightmare scenarios in which a standard government-approved education was provided to every citizen in pill form, or in which only an elite class was provided with certain pills.

✧   But why would such nightmare scenarios be any more likely if knowledge were pill-based than they are now?

✦   It is an accepted function of government to control the kinds of drugs its citizens are allowed to consume, based on concerns for public health and safety. Some countries also closely regulate the kind of education their citizens receive, often through national standards, but such regulation could not compare with what would become possible if knowledge became pill-based.

✧   Let's assume the government that designed and distributed the pills had the best of intentions. What would be the harm in that case?

✦   Even assuming benign intentions, the acquisition of knowledge through pills would have many bad effects. For one, it might well lead to a conformity in our thought processes that would make it much more difficult for new ideas to come forth. For another, it would completely undercut our incentive to expend any mental effort to learn things the old-fashioned way. Our mental abilities might atrophy, and we might depend on Big Brother to install our brain's "software."

What if porpoises were more intelligent than people?

✧   Are you implying that in the real world this might just be true? And why did you pick on porpoises?

✦   Porpoises make a more interesting comparison species than, say dogs or elephants, because their very different anatomy and environment makes intelligence comparisons much trickier.[1] If dogs were more intelligent than people,

---

1. Here's an IQ test to see how your IQ compares to that of an elephant. How might you (as an Indian elephant) sneak into your owner's plantation at night to steal bananas without ringing the bells placed around your neck? Answer: elephants have learned to place mud inside the bells.

they would have to do a fantastic job of hiding that fact. On the other hand, it is not out of the question that porpoises are, in some sense, more intelligent than people. In fact, it seems quite likely that the relative ranking of the intelligence of various species would depend greatly on which species got to design the IQ test.

✧ Maybe we should allow the porpoises to make up the test, and see how we do on it.

✦ For all we know, they may already have given us an IQ test, and found us to be "moderately" interesting creatures, second only to themselves.

✧ But aren't there objective criteria, such as brain size, that could be used to judge intelligence?

✦ It's probably more relevant to look at brain weight as a fraction of body weight. Using that measure, humans come out quite well, but apes and porpoises come out nearly as well.

✧ OK, how about the ability of a species to create tools, modify its environment, and build complex structures? Surely that shows the superiority of humans over other species.

✦ Not really. You picked accomplishments that you knew humans were better at than other species. Different groups of humans show a vastly different degree of interest in such things. Many indigenous peoples, for example, show little interest in modifying their environment and building complex structures, and much more interest in living in harmony with their environment.

✧ I guess if the IQ test included questions on such things as the night sky, the ways of animals, agriculture, and herbal medicine, most Westerners would do very poorly compared with indigenous peoples. But shouldn't porpoises have built something if they are intelligent?

✦ Not necessarily. First, remember they don't have any hands. Secondly, porpoises have no need for many of the things that stimulate human construction activities: they don't need shelter, agriculture, clothing, fire, or transportation vehicles, and they may be smart enough to realize they can do without weapons. In fact, some creatures such as ants or bees that engage in very extensive construction activities are not rated as being particularly intelligent—such construction being attributed to instinct.

✧ Well, if we cannot accurately judge porpoise intelligence based on brain weight or interest in complex construction, what about the degree to which porpoises exhibit complex behavior or complex language?

✦ Porpoises do show a rich range of behaviors, although we cannot really know for sure the purpose of much of it. And as far as language is concerned, we know nothing regarding the meaning of porpoise sounds, or for that matter the rich array of whale sounds. Isn't it interesting that we tend to view animals such as apes as intelligent to the extent that they understand our language, not by how well we understand theirs?

✧ How about the fact that porpoises and people seem to get along so well?

✦ Well, people and dogs also get along very well, but with porpoises the relationship seems more one of near equals than master and pet. But if porpoises are in fact more intelligent than people, they might be content to continue with their own pursuits, rather than to impress us with their intelligence and raise the specter of how "useful" they might be to us.[2]

✧ I suppose it's unlikely they would want to recruit us for their "Land World" show.

What if everyone knew the date of their death?

✧ I think it would be terrible to know just when my number would be up.

✦ You might feel differently, depending on just when your death date would be. If you were told it would be fifty years from now, you'd probably feel a lot better than if you were told it would be next week.

✧ Even with fifty years to go, every day I woke up I'd think, OK, X more days to go, and then the END.

✦ Someone else might say upon waking: another day that I don't need to fear dying.

✧ True. But those two kinds of reactions are not so different from the ways people now react to life. In any case, why would anyone want to know their death date?

✦ Financially, it might appear as if there were many advantages. For example, you could stop worrying about whether you would outlive your savings in your old age.

✧ Why do you say "might appear?"

✦ Because it would still be possible to get sick or badly hurt, so you would still need to save money for such contingencies.

✧ Would you live your life any differently?

✦ I'm not sure. Some people might be willing to take all kinds of risks knowing they wouldn't die, but again the possibility of getting badly hurt should make

---

2. Much evidence about the intelligence of porpoises is anecdotal. For example, the astronomer Carl Sagan has noted that in once playing with a porpoise, he had the distinct impression that the creature was testing his (Sagan's) intelligence and reactions.

you just as cautious as you would normally be. The biggest difference is that now people can effectively put death out of their minds and live their lives without thinking about it—which might be difficult to do if they knew the date of their death.

✧ How could nature possibly arrange things so that people knew the date of their death?

✦ If we get into the realm of the supernatural, God could communicate this information to everyone. But staying within the natural realm, it would seem possible for nature to determine only your "normal" expiration date, that is, the date your body would wear out, assuming you didn't die first due to accidents or other causes.

✧ How might that work?

✦ The only way this could happen would be if our bodies were much more standard than they are now. There would need to be some part of our bodies that got "used up" at the same rate in everyone. Even for manufactured items there are great uncertainties as to when something will wear out.

✧ About the only thing you can count on is that things will wear out right after the warranty expires.

✦ One hypothetical possibility is that there is some substance initially present in your body in a specific amount. Let's call it your level of "zorgone." Suppose that as you lived your life you used "zorgone" up at a constant rate each day, independent of your activity level. In that case, your life span would be known exactly based on your initial zorgone level.

✧ Could there actually be such a substance in our bodies?

✦ No, I just made it up, because I couldn't come up with any other way nature could arrange to give everyone a specific, fixed life span.

✧ I think I'd quit working and have fun if I knew I were going to die next year.

✦ As they say, "Life is short . . . eat dessert first." We should all live our lives as though today might be our last one. That shouldn't keep us from making plans for the future; it means that we need to learn to enjoy each moment.

What if people lived a thousand years?

✧ I imagine I could no longer collect social security at age 65. But how could such a thing be possible? We cannot stop the aging process, can we?

✦ The aging process occurs in our bodies at the cellular level. Every time cells in a mature person divide, little bits at the ends of the chromosomes get lost, leading to some degradation each time, as the body keeps replacing itself.

✧　Too bad the chromosomes don't have those little things at their ends like shoelaces have to keep their ends from fraying.

✦　Actually, some cells in our body do just that, in effect, by creating a certain enzyme that stabilizes the ends of the chromosomes and prevents them from breaking off during cell division. These are the sperm and egg cells produced by the sex organs. These cells need to be "immortal" in order to faithfully transmit the "recipe" for making life to the next generation.

✧　Too bad other cells in our bodies can't also make that same enzyme and be "immortal."

✦　The only other cells that now fall into that category are cancer cells, whose immortality often causes them to spread unchecked—stopped sometimes only at the cost of our own mortality. So it is possible that if we somehow got other cells in the body to emulate cancer cells and manufacture that special enzyme, they too might become immortal.

✧　Either that, or else everyone would get cancer. Let's think of another way to achieve a thousand-year life span.

✦　In the far distant future, it might be possible to extend the idea of organ transplants to whole-body transplants, although the idea is probably unfeasible given ethical and practical constraints, such as the unavailability of donors.

✧　What about the half-human, half-machine androids of science fiction movies? Why not transplant human brains into mechanical bodies? Or to go to the logical next step, we can imagine all of a person's memories and thoughts put into the mechanical brain of a robot.

✦　Of course, in the science fiction movies they can never transplant the soul as well, and very bad things happen—prompting the conclusion that we should not tamper with the forces of nature.

✧　And that scientists are a bunch of nerdy evil maniacs! But what would our society be like if we actually did have a thousand-year life span?

✦　Presumably, since most people tend to become more conservative as they get older, if we did live a thousand years, it seems likely we would be less willing to take risks and would become creatures of habit to a much greater extent than now.

✧　I guess society would become much more politically conservative.

✦　But not just politically. With a thousand-year life span, the main causes of death would probably be accidents and possibly murders and suicides, so we would be much more cautious in our everyday lives.

✧　But why would the accident and murder rates increase just because people lived much longer lives?

◆ People eventually have to die of something. The only way that a life span of a thousand years would be possible would be if the effects of disease, especially the disease we call old age, were largely eliminated. In that case, other causes of death would assume greater relative importance.[3]

✧ With a thousand-year life span, I guess there wouldn't be much room for new children. You could even imagine having strict rules limiting who could have kids and how many they could have, as is the case in China now.

◆ Right. And it's not so clear that some people would be able to live a thousand-years without becoming bored out of their minds.

✧ Hooray for death!

What if you could go into a state of suspended animation?

✧ By suspended animation, do you mean being asleep for an extended period of time? That could be a nice way to get through some boring dates I've had.

◆ Suspended animation is more than a state of sleep, since most bodily functions would effectively be put "on hold." The closest nature comes to this state is when bears and some other creatures hibernate for the winter.

✧ Do we know how bears hibernate?

◆ Not really. We do know that before hibernation bears put on about a hundred extra pounds of fat, which they live on during the winter months while they sleep—dozing off for weeks at a stretch. In fact, female bears even give birth while they are asleep.

✧ How painless! I don't want to be indelicate, but what about some of the usual bodily functions bears and humans have? Wouldn't a bear at the end of its hibernation find itself in a somewhat messy state?

◆ We don't know how, but the bear's body breaks down wastes internally, converting them back to protein.

✧ What would be some uses of suspended animation if we learned how to do it?

◆ One obvious use would be for space travel over long distances. A voyage

---

3. For example, currently your chance of being a victim of a murder or suicide is roughly 0.02 percent per year. If the rates stayed the same as now and people had a thousand-year life span, the chances of your being a suicide or murder victim during a thousand-year period would be roughly 20 percent.

to Jupiter, for example, would take several years with present-day rockets, so suspended animation would be an attractive possibility.

✧ But I don't think I'd want to have a friendly computer named Hal looking after things while I snoozed. What other uses might there be for suspended animation?

✦ Some curious types might want to see how the whole human drama plays out in the future. Like a time traveler, they might wish to go into suspended animation for years, in effect fast-forwarding through future history and waking to observe what happens. Of course, there are some important differences between this type of time travel and the real thing.

✧ Like what?

✦ First, if you went into suspended animation by, say, freezing your body, the trip to the future would be strictly one way, so if you didn't like the way things turned out in the future, there would be no going back. Second, if things in the future turned out really badly, there might be no one to release you from your suspended animation. Or, even less dramatically, the company with its freezer full of bodies may have gone out of business and someone may have pulled the plug—which actually has happened in at least one case. But space travel and time travel are probably not the main reasons many people would want to go into a state of suspended animation.

✧ What do you think the main motivation would be?

✦ Some people dying of illnesses that are incurable today might choose to have their bodies frozen to be defrosted at a time when scientists have learned how to cure those diseases. This procedure is now legal only after people die, so presumably scientists would also need to figure out how to bring people back to life.

✧ But suppose people were frozen just prior to their death. Would that make the job of future scientists a lot easier?

✦ Not a whole lot. The main problem seems to be in the defrosting process, during which a lot of cell damage would occur. But there is a species of frog that freezes each winter and defrosts in the spring with no ill effects.

✧ The business of keeping a bunch of dead bodies frozen for future scientists to resurrect sounds like something out of a horror movie, and anyway it must be quite expensive.

✦ The outfit that does it supposedly has a special rate on "head jobs." Many people opt just to have their heads or other favorite parts frozen in the hope that future scientists will be able to outfit them with artificial bodies.

✧ I guess many people would like to cling to any possibility, however remote, of becoming immortal. I think I'll just have a few strands of my DNA preserved and let them recreate me, as they did with dinosaurs in *Jurassic Park*. Or, better yet, why not just have kids?

What if two or more people could time-share a body?

✧ By time-sharing I assume you mean inhabiting the same body at different times—as in the real estate concept of time-shared ownership?

✦ Yes, but let's also consider simultaneous occupancy as well.

✧ I understand that some people who have been hypnotized and supposedly regressed to past lives claim to have time-shared the same body.

✦ I suppose, if reincarnation were possible, that would only be natural, considering how many more people are alive now than in the past. In fact, if we project back to the early days of mankind, each person then would have had to have been host to literally millions of occupants—assuming everyone alive today has been reincarnated many times, and always in human form.

✧ Are there more plausible examples of "time-sharing" other than reincarnation?

✦ People diagnosed as having multiple personalities could be said to live a time-shared existence, with different personalities taking occupancy at a given time.

✧ If the different personalities are not aware of the actions of the "previous tenant," it must lead to some bizarre sorts of situations—as seen, for example, in the plots of many a murder mystery.

✦ Even though multiple personality disorders may be rare, in a certain sense each of us is in the position of having a time-shared existence. Scientists have established that different parts of our brains are the sites from which different types of thoughts and feelings arise. It seems that our brains consist of a number of subbrains linked together internally.

✧ What do you mean by subbrains?

✦ It is well established that the left and right hemispheres of the brain handle somewhat different kinds of activities. Language, for example, is primarily the province of the left brain, whereas tasks such as understanding spatial relations are primarily handled by the right hemisphere. To greatly oversimplify, we can say that our left brain is the logical part and our right brain the intuitive part.

✧ How can the connection be made between particular structures in the brain and thoughts, perceptions, and feelings?

✦   This connection has been determined largely through the electrical activity picked up by many electrodes placed around the skull of a subject. Another important source of information has been from people who have had a particular part of the brain damaged, and who lack certain specific abilities as a result. One particular group of people who have been studied extensively are split-brain patients.

✧   You surely can't mean their brains have been split?

✦   Yes, actually, they have been. In certain people who have violent epileptic seizures, scientists have found that the seizures can be relieved by severing the connecting link between the left and right hemispheres, the so-called corpus callosum.

✧   How do these people behave—as if they had a split personality?

✦   If you met one, you probably wouldn't notice anything strange, unless of course that person was strange before the operation. But careful tests show that split-brain patients indeed behave as though they were two individuals occupying the same body, one of which receives certain types of information from the environment, which the other one learns about only secondhand—*by experiencing how the body reacts to this information!*

✧   So I suppose we could say that body time sharing goes on with all of us now, but the communication between the "tenants" is normally good enough so that we usually have the sense of being a single individual.

What if you slept every other minute?

✧   There isn't anything comparable to this kind of sleep in reality, is there?

✦   Not every other minute, but many insomniacs who get little or no sleep at night have many very short "microsleeps" during waking hours. Apparently, the body needs a certain amount of sleep and will manage to get it, one way or the other.

✧   What's the point of sleep, anyway? It seems as though evolution would favor creatures that didn't need to sleep, and wouldn't fall prey to other creatures while asleep.

✦   By the same reasoning you might say that evolution would favor creatures that didn't have to eat! Eating, of course, is needed to resupply cells with nutrients used to generate energy. In the same way, brain cells need to generate electrical currents. They are like electrical batteries that generate electricity through chemical reactions.

✧ So the commonsense idea that sleep is nature's way of "recharging our batteries" may be more than just an analogy.

✦ That's right, although unlike calculators, where the batteries are in a separate compartment, in the brain the energy is stored chemically throughout the organ on the molecular level.

✧ But why do most people sleep once a day? Our muscles, which also need to be chemically recharged when they are overused, can apparently be recharged on a continuous basis.

✦ One might imagine that mammals developed with a variety of possible sleeping arrangements. Two that have apparently been favored by evolution are sleeping during each night and sleeping during each day. We might imagine creatures that slept, say, one day every four, but perhaps the brain needs more frequent "recharging." Other sleep arrangements don't seem very promising either. For example, sleeping every other minute could be extremely dangerous to a creature engaged in activities lasting longer than a minute!

✧ How about alternating sleep and wakefulness even more frequently than once per minute? Why not take things to a ridiculous extreme and imagine that we sleep one-hundredth of a second out of every three. That way, during the course of each day, we would manage to sleep a total of eight hours.

✦ That's an intriguing possibility. It potentially gets around the problem with sleeping every other minute, in that you might not be continually blacking out. Consider the analogy of driving a car over a road with many gaps. If the gaps were a few feet across, you couldn't drive on the road without continually falling into them. But with, say, quarter-inch gaps only an inch apart, you wouldn't even notice that there were any gaps while driving, much as blinking doesn't usually interfere with seeing.

✧ Are you suggesting that some creatures actually sleep this way?

✦ There is no evidence one way or the other, but it could explain those exceptional people who seem to get by with next to no sleep. Of course, it is also possible that these people have unconsciously mastered the trick of "recharging" their brain while it is being used, in much the same way that the alternator continually recharges the battery of your car even while the battery supplies current.

✧ That seems much preferable to our present once-a-day sleeping arrangement. Why wouldn't evolution have favored animals capable of this?

✦ Your guess is as good as mine. One possibility is that sleep keeps your body still, and less noticeable to nighttime predators. Then again, sleeping one-hundredth of a second every three might be equivalent to operating mentally at 67 percent efficiency. People might be better off if they operated at 100 percent efficiency during their waking hours.

What if you had a near death experience?

✧ I assume you mean "seeing the light" and all that.

✦ Yes. Many people who have been very close to death, some with their heart stopped, have reported such experiences. According to a psychiatrist, Dr. Raymond Moody, the components of this experience include hearing the news of one's death, having feelings of peace and tranquillity, entering a dark tunnel, leaving one's body, meeting dead relatives, seeing a bright light, being spoken to by a being of light, experiencing a review of one's life, and coming back to one's body.

✧ Do most people who have been near death report these experiences?

✦ According to one study by Moody, roughly half the people interviewed reported some form of a near death experience (NDE), but it usually included only a few of the elements listed. For example, only about a third reported some sensation of body separation, and only a sixth reported seeing the light.

✧ How do we know this is not some kind of dream or hallucination? I would expect religious people, or those who have read about NDEs to fantasize along these lines, if they knew they were near death.

✦ We cannot rule out the possibility of the experience being a dream or hallucination, but it is curious that there are so many common elements and so few hellish experiences. According to one study, there was no connection between the strength of one's religious beliefs and having a NDE, and those who had previously read about the subject were not necessarily more prone to have one.

✧ Is there any way to explain NDEs based on what we know of the physiology of the brain?

✦ There have been several partially successful attempts. When the brain suffers oxygen deprivation, which would be the case when the heart stops pumping blood, the physiological condition induces perceptions very similar to those of the NDE—especially the dark tunnel and the light. In addition, if a certain region of the brain (the sylvan fissure) is directly stimulated by electrodes, the subject also experiences some of the features of an NDE, including a feeling of leaving one's body.

✧ But can neurology explain all the features of an NDE?

✦ It is not necessary that there be a single explanation of all parts of the NDE. Perhaps lack of oxygen induces the relevant parts of the brain to be stimulated, along with the optic nerve (seeing the light), which in turn induces a hallucination.

✧ Haven't there been some cases where the person having an NDE sees his

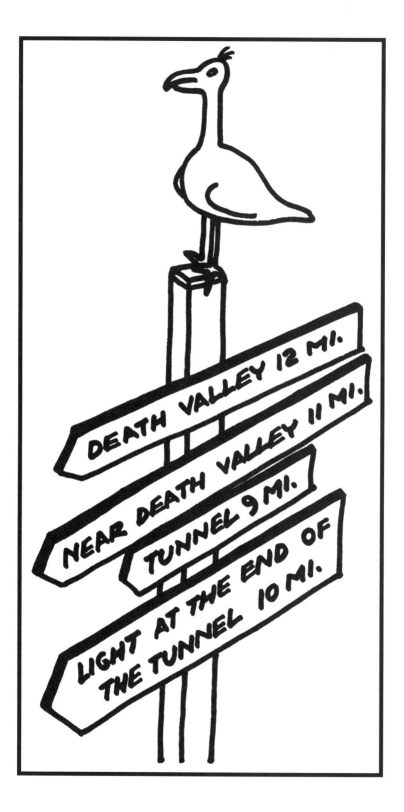

own body from a height and can afterward report many specific details of what was said and done while he was near death?

✦ There have been a handful of such cases, but it is difficult to confirm their authenticity. Also, it is quite possible that in some cases people in comas can actually perceive what is going on around them.

✧ So are you saying that NDEs clearly have a physiological explanation, or is there some way they could be shown to represent a real experience—one that strongly implies the existence of an afterlife?

✦ We can never prove that NDEs are due merely to physiological experiences, even if physiological experiences closely mimic them. But there is a possible way to show that they represent something beyond the physiological realm. One researcher has placed a flashing electronic sign in a hospital emergency room at a location that could be seen only by someone who left his or her body and hovered near the ceiling. The message on the sign is changed frequently, and obvious messages like "What are you doing up here?" are avoided. If some NDEers in a number of different hospitals ever should correctly report the content of these messages while supposedly out of their bodies, this would be highly persuasive evidence for the reality of the experience.

✧ Suppose we get back to our original question: what if *you* had a near death experience? How would it affect your life?

✦ I imagine my reaction would be similar to that of other people who have had such experiences. Many people who have had NDEs report profound changes in their outlook on life. Most importantly, armed with what they perceive as a firsthand preview of an afterlife, they no longer fear death, and they conduct their lives with a renewed sense of purpose.

What if you stepped into a "matter transporter"?

✧ "Beam me up, Scottie. There's no intelligent life down here." You mean that type of matter transporter?

✦ Yes. Let's imagine that a device were built that could disassemble the molecules of your body and transmit a signal that would permit your body to be exactly replicated at some remote point, just as on *Star Trek*. Would you be willing to step into the device?

✧ Well, I certainly wouldn't want to be an early pioneer in its use, but if I saw many people successfully transported without incident, I'd be willing to do it.

✦ But remember that the original you that steps into the machine at one end is obliterated. It is only a copy—albeit an exact copy—that comes out at the other end. The copy will have all your memories and will appear to others as you would, but it will not be the you that stepped into the other end. It will not have the same consciousness that you have, even if it has the same memories.

✧ But why are you assuming that the being coming out the other end is only a copy, and not the original me?

✦ Because it is assembled from a different collection of atoms. It would be much like following a blueprint to assemble an identical copy of a piece of machinery.

Beam us up, Scottie; there's no ~~babes~~ down here.
intelligent life

✧ Are you saying that the duplicate being that is created at the other end is just like me but without a soul?

✦ We're talking about consciousness here—your conscious awareness that wakes up every morning to find itself in the same body, not someone else's. The duplicate of you at the other end of the machine will presumably have its own consciousness, but it just won't be yours. To make this clearer, let's say the machine were to transmit a copy of you but not destroy the original in the process. Clearly, the awareness you possessed would continue to reside in the original you.

✧ If the duplicate of me had the same memories as I do and thought just like me, might it claim to be the real me? In science fiction stories of this type, the duplicate usually has some sinister character flaw that allows outside parties to identify the "real" original. But for an exact duplicate, no such identification would be possible.

✦ Yes, but whatever the duplicate claimed, you (the original) would know that your consciousness resided in *your* mind.

✧ This whole discussion is science fiction, so I don't know how you can be so sure that my double wouldn't have the same conscious awareness I do. Can I assume there's no way *you* would step into the matter transporter?

✦ Only under one condition: If people who stepped in one end found that they and their duplicate at the other end had a shared consciousness. In other words, whatever one of you saw, thought, and felt was shared by the other.

✧ Such a thing isn't really possible, is it?

✦ Who knows? We do know that sometimes identical twins seem to have a direct awareness of one another's thoughts. Conceivably, you and your exact duplicate could have, in effect, a shared consciousness—although, frankly, the idea does seem somewhat implausible, because each of you would receive different sensory inputs. Perhaps the situation for the two of you would be akin to the two hemispheres of your brain, both of which have a shared consciousness.

✧ Barring the possibility of a shared consciousness, what is this mysterious conscious awareness, which would reside with the original me and not with my exact duplicate (whatever he might claim to the contrary)?

✦ That's a question thinkers have wrestled with for centuries. Despite all the advances of science, we are no closer to answering that question than before.

What if radiation were good for you?

✧ I hope you're not going to tell me about the benefit of having a few extra fingers. Hasn't it been amply demonstrated that radiation is harmful to living things?

✦ High doses of radiation are unquestionably harmful—sometimes lethally so. Radiation is a mutagen (cause of genetic mutations) as well as a carcinogen (cancer-causing agent). Of course, radiation also is used in treating cancer, since focused beams of radiation can target and destroy a tumor while doing less harm to the surrounding tissue.

✧ But what about the rest of us who don't have cancer? Are you claiming that some radiation might be good for us too?

✦ Radiation is clearly beneficial as a diagnostic medical tool. A typical chest X-ray, for example, might expose you to only a thirtieth as much radiation as you would receive annually just from living on planet Earth. The potential harm from this addition to your normal annual dose is minuscule when expressed as an increase in your risk of dying of cancer, and it is far outweighed by the potential benefit of early cancer diagnosis.

✧ OK, I understand that radiation does have its beneficial uses, but what about the idea that radiation might actually enhance your well-being apart from these uses?

✦ At one time people did actually believe this. In fact, you can still find places advertising their "radium hot springs" baths, and the idea is still considered credible in some Eastern countries. At present, however, it is generally assumed that any exposure to radiation carries some risk, and that the level of risk is proportional to the dose you receive. This assumption is called the "no-threshold model," because it assumes that there is no dose below which no harm occurs.

✧ Why do you say the no-threshold model is generally "assumed"? Isn't there good empirical evidence that the harm done by radiation is, in fact, proportional to the dose received?

✦ Such a direct proportionality has been found to hold true for high doses, both in animal studies and in studies done on survivors of the atomic bombings of Japan during World War II. The difficulty is that, given statistical uncertainties in the data, for lower doses we have no way to tell whether or not any harm is done by radiation. Let's take an example. If we assume that the harm done is strictly proportional to the dose (i.e., the no-threshold model), your risk of dying from cancer from a single chest X-ray would be around 0.0002 percent. In other words, your lifetime risk of dying from cancer might go from around 20 percent to 20.0002 percent.

✧ Putting it that way, a single chest X-ray doesn't seem like much to worry about. But why can't scientists tell if that 0.0002 percent increase in risk is real?

✦ Epidemiologists who study such things can look at the numbers of cancer deaths in groups of people who have received various estimated radiation doses.

But the random variations, and the possible variations due to other causes of cancer, make it impossible to see if such a tiny increase in risk is real—no more than a pollster could say one candidate was really ahead who had a 0.0002 percent lead in the polls. In fact, for all we know, radiation in small doses might actually be good for you—a hypothesis known as hormesis.

✧ Ah yes, I can just see the ads now: "Plutonium Tanning Salons—you'll leave with a glowing, tingling feeling." Surely you can't be serious.

✦ Actually, many substances are harmful to humans at high doses and beneficial at low doses. Some examples include fat, vitamins, iodine, aspirin, zinc, iron, aluminum, sunshine, and possibly even alcohol. It seems quite conceivable that carcinogenic substances in very small doses might stimulate the immune system so as to lower your cancer risk. The effect would be akin to that of vaccines, which create immunity to a disease by giving you a mild case of it.

✧ But wouldn't there be some empirical evidence if low doses of radiation were actually beneficial?

✦ T. Don Luckey, a professor emeritus at the University of Missouri, claims in his book *Radiation Hormesis* to have found many instances supporting the hormesis hypothesis. One of them involved women who received radiation in the form of fluoroscope exams during treatment for tuberculosis. Those women who received very low doses appeared to have a lower death rate from cancer than the general population. Another scientist, Dr. Bernard Cohen, has likewise found a *negative* correlation between lung cancer and radon levels in homes— higher radon levels corresponded with lower lung cancer rates. But most other scientists remain highly skeptical of such claims. Usually, either the claimed effect is of dubious statistical validity, or else the possibility of confounding variables cannot be ruled out.[4]

✧ But if we don't really know for sure whether very low doses of radiation are harmful or beneficial, isn't the simplest and safest course to adopt the no-threshold view, which assumes that the level of harm is strictly proportional to the dose received?

✦ It may be simplest in a mathematical sense, but it isn't necessarily the safest course in terms of public policy. For example, the public has become so convinced of the need to eliminate even trivial radiation exposures that we have embarked on a cleanup of nuclear weapons plants that may eventually cost $200 billion. The end result of that cleanup may have zero or no more than a minimal impact on improving our health.

----

4. In a recent paper, the author has shown that Cohen's strange result can be explained by the confounding effects of smoking, that is, those countries having higher radon levels tend to have slightly lower percentages of smokers.

✧   Are you advocating abandonment of such cleanups, or perhaps adding new nuclear weapons plants to ensure that everyone gets their desirable allotment of radiation?

✦   No. But if we had a more realistic view of the magnitude of the danger, we might find solutions that cost far less than $200 billion, which, if invested in realistic ways to improve the nation's health, could have enormously greater impact.

Chapter 6

Perception

What if you heard sounds only when you faced the source?

✧   I thought that many animals do have directional hearing that allows them to tell from what direction a sound comes.

✦   Yes, and people have that capability to a limited degree as well. But here we're considering the possibility that you wouldn't even hear the sound unless you faced the source.

✧   Since we see things only when we are looking directly at them, it does seem strange that we can hear things even though our back may be turned.

✦   Actually, not all sounds can be heard equally clearly when your back is to the source. High-pitched sounds, like those in a whisper, are difficult to hear in this case. Have a friend whisper something and turn your back after he starts, and you'll see a big difference.

✧   So that's why it's considered sneaky to whisper behind someone's back. But why should high- and low-pitched sounds behave so differently?

✦   Sounds travel through a medium like air or water in the form of waves. Waves with a long wavelength bend around things, but waves having a short wavelength, such as light, tend to travel in straight lines we call rays. High-pitched sounds have a shorter wavelength than low-pitched sounds, and so they bend around obstacles like your head or the back of your ears less than low-pitched sounds.

✧   So that's why I've heard of light rays, but never sound rays.

✦   Actually, there is a situation where sound travels in straight lines or "rays." This happens any time the wavelength of sound is small compared with the sizes of objects the waves encounter.

✧   I'm not sure I follow this. Could you give me a simple example?

✦   Sure. If you were in a very long tunnel (say 50 or more feet from one end), a friend just outside the end of the tunnel could hear you shout. But if she moved to the side, where she could not see into the tunnel, she would hear you much fainter, if at all. On leaving the tunnel, the sound of your voice would mostly be beamed in the forward direction until it encountered other objects.

✧ So, in this example, the sound from my voice really could be said to travel in straight-line "rays," because the wavelength is much less than the width of the tunnel, right?

◆ Exactly. Perhaps the best example of "hearing" sounds only when facing the source would be the directional microphone. These are made from a collection of small microphones all mounted along a rod. When the rod is pointed at the source of sound, the signal received by the microphone is much louder because of the way the separate signals are combined.

✧ So what would life be like if sound always traveled in straight-line "rays," and we only heard sounds when we faced the source—just like light, or just like a directional mike?

◆ The situation with regard to human speech would be analogous to a deaf person with excellent lip-reading skills who could understand speech, but only when looking at the speaker. Such highly directional hearing would make us unaware of distant events on a 360-degree basis—an ability furnished by no other sense. Even our sense of smell doesn't have a 360-degree capability, since we can only smell things that are downwind. The absence of a 360-degree "early warning system" could be fatal in an environment with a lot of predators around.

✧ So how would nature have handled the situation?

◆ Probably by one of two solutions: either by having animals evolve with the ability to swivel their ears to receive sounds from different directions—an ability that some animals now do have to a limited extent—or else through the evolution of a couple of additional ears (or eyes) pointing in different directions.

✧ Why eyes?

◆ Well, we don't need eyes in the back of our head precisely because our ears can now pick up sounds from all directions. But if that weren't the case, an extra eye in the back of our heads might come in mighty handy.

✧ Especially for parents with small children.

What if you could "see" sounds?

✧ Yeah, every time I hear the song "Ninety-nine bottles of beer on the wall," I see a picture of my old college fraternity.

◆ Actually, I had something else in mind, namely, being able to form images of sounds directly, which apparently is the case with some people who perceive a certain image or color every time they hear a certain sound.

✧ Being able to see sounds seems as nonsensical as "hearing" odors or pictures. Isn't each of our senses tied to a particular way of receiving information from the outside world?

✦ That's true. In fact, the difference between sight and hearing is especially profound dimensionally. With each eye we perceive images in two spatial dimensions, but with each ear we perceive only the dimensions of loudness (intensity) and time (duration).

✧ But I know I've seen pictures of sound produced by special devices, and we can also get directional information from sounds, based on the stereo effect of having two ears.

✦ One way of making "sound pictures" or sonograms is to convert sound into images by mapping the time dimension on one axis of a graph and loudness on the other. But these are visual representations of sound rather than a direct perception. Likewise, while the directional information provided by stereo sound provides some degree of two-dimensionality, it is a far cry from an image.

✧ How about those sonograms of fetuses in the mother's womb taken with ultrasound—would that qualify as seeing sound? And what is ultrasound anyway?

✦ Now we're getting close to what I had in mind by "seeing" sound. Ultrasound is simply sound that is at too high a pitch or frequency to be audible to the human ear. Its high frequency and short wavelength means that in a uniform medium, ultrasound tends to travel in straight-line "rays" just as light does. As a result, ultrasound, like light, can be focused using lenses, and images can be produced using special cameras.

✧ But it's one thing to produce images with special devices and something else to be able to see sound directly, isn't it?

✦ That's true, but consider the bat. Bats emit continuous ultrasonic squeaks, and they continually monitor their environment based on timing the echoes. A bat's information about its environment sensed from these ultrasonic echoes is so good that bats can fly in complete darkness, even navigating so skillfully as to avoid wires strung up to test their skills. We can speculate that a bat uses ultrasound to produce direct images of its environment in much the same way that we use light. The bat would then literally be seeing its environment in sound.

✧ How about blind people, who tend to have a highly developed sense of hearing? Don't they do more or less the same thing?

✦ When blind people tap their canes, they do get information about their environment in a similar way to bats. But a blind person probably gets a far poorer picture this way than a bat can, for a number of reasons. First, bats use ultrasound, which can form true images because it has a much shorter wavelength than the sound we hear. Second, bats emit a continuous rapid stream of clicks, which allows them to continually update these images.

✧ Why doesn't some clever engineer build a device for blind people that emits a steady stream of ultrasonic clicks and converts their reflection to an audible signal? Maybe blind people using the device would be able to form images of their environment just like bats.

✦ That's an interesting idea, but it probably wouldn't work, because bats' brains are probably "hardwired" to interpret their environment through sound. In contrast, humans, equipped by evolution to rely primarily on sight, might not be able to acquire such a skill. But some clever engineer should give it a try anyway.

What if we had a sixth, seventh, or eighth sense?

✧ Would that be in addition to the five we have now: sight, hearing, smell, touch, and taste?

✦ Right. By some counts, though, we already have more than five, if you count our senses of balance, temperature, and pain. In addition, you could say that we have a direct sense of time, based on our internal "biological clocks." Some people reportedly have such a good sense of time that they can tell you the elapsed time to the nearest second after a period of many minutes.

✧ How about our ability to sense motion, and let's not forget our sense of direction and our sense of humor?

✦ Actually, we do *not* have an ability to sense our body's motion through space if that motion is at constant speed in a straight line. Accelerations and decelerations are another story, as you know from riding elevators. As far as humor and direction are concerned, they are not true senses with their own receptor organs. (I don't count the funny bone.)

✧ How about the "sixth sense" of extrasensory perception (ESP), including clairvoyance?

✦ Somehow I knew you were going to bring that up. For now, let's stay with senses that are connected to known sense organs and receive some kind of classifiable signal. If there is anything to ESP, it presumably would have to have a physical basis, even though we might not know what it is.

✧ When considering possible additional senses, I think we want to go beyond things like ultrasonic sound heard by animals, since that kind of signal is basically just sound of a higher frequency.

✦ Right. We are after totally different "windows" on the environment. A look at the animal kingdom, however, does suggest some possibilities. Electric eels, for example, not only act as a source of electricity, but they also can directly sense their electrical environment. Other creatures, including birds, seem to have a magnetic sense, which provides them with a built-in compass for navigation.

✧ How do we know that birds navigate based on magnetism?

✦ We can try to fool their magnetic sense by setting up magnetic fields to mask Earth's field and seeing how this affects the birds' sense of direction. Actually,

A useful sixth sense.

however, birds appear to have a couple of other navigational aids besides magnetism. They seem to be able to directly sense the motion of the sun and the stars.

✧ Do people have either of these senses to any degree?

✦ Some people do seem to have a much better sense of direction than others, and some (especially men) will do anything to avoid facing the fact they are lost. But there is no evidence that the human sense of direction is based on a direct perception of either Earth's magnetism or the motion of the stars.

✧ How about some other possible senses not present in the animal kingdom?

✦ Nature seems to have been very resourceful in providing its creatures with sensory receptors for coping with most aspects of the natural environment that can affect them. One that appears to have been overlooked is a sense organ to perceive radioactivity. Perhaps if we lived in a world where radioactivity levels varied appreciably from place to place and there were a big advantage in being aware of "hot spots," such a direct perception might well have evolved.

What if your reflexes were much faster or slower?

✧ What exactly do we mean by reflexes or reaction time, anyway?

✦ Your reflexes or your reaction time would be the time that elapses between your seeing the bad guy go for his gun and the instant when you start to go for yours. Actually, having faster reflexes might not allow you to draw a gun much faster, since an equally important limitation is how fast you can move your hands.

✧ I'm not sure I see the distinction you're making.

✦ A more extreme case would be if you lost the draw and wanted to dodge the bad guy's bullet. Even if you had infinitely fast reflexes, you couldn't do it, because your body could not move fast enough to get out of the path of the bullet, even if you started moving at the instant you saw the bullet leave the gun.

✧ What determines how fast my body can move?

✦ It ultimately comes down to muscle power. Sprinters have much more muscular thighs than distance runners so as to accelerate off the starting blocks with the greatest acceleration. For short races (and for dodging bullets), acceleration is much more important than speed.

✧ What is the average person's reaction time?

✦ About one or two tenths of a second. Here's a simple way to measure yours. Have a friend drop a ruler between your outstretched thumb and index finger, and try to grab it between your fingers as soon as you see it fall. With a reaction time of a tenth of a second, you'd be able to grab it after it fell 2 inches, and with a reaction time of two-tenths of a second, it would fall nearly 8 inches.

✧ So what is it that limits my reaction time? Is it something I can improve by practice?

✦ Your reaction time is set mainly by the speed at which electrical nerve impulses travel from your brain to your extremities—about 10 meters per second. A sprinter who, through practice, gets off the starting blocks faster than her reaction time is not reacting to the sound of the starter's gun but instead to her anticipation of the standard time interval between the announced "ready . . . set . . . bang!"

✧ What would life be like if our reflexes were much slower than they are now?

✦ You might get poked in the eye a lot if your involuntary reflex to close your eyelid when a branch approached didn't work as fast. Also, a lot of activities, like driving a car, would become much more hazardous. Your ability to drive would be impaired like that of a drunk. In fact, with very much impaired reflexes, you'd probably fail a sobriety test for walking a straight line.

✧ What does walking a straight line have to do with the speed of my reflexes?

✦ Walking is a very complex activity involving the transfer of weight from leg to leg in a precise time sequence. If you transfer the weight a little too soon or too late, you will be very unsteady. In fact, with really slow reflexes, you probably wouldn't even be able to stand erect.

✧ You'd better explain that one.

✦ When standing, your body is inevitably swaying to a small extent. To keep from toppling over as your weight shifts, you need to sense your body's motion and adjust the pressure on your toes to compensate before the motion progresses too far in one direction. With very slow reflexes (or very small feet), standing would be impossible.

✧ I guess I should be thankful that I have big feet to make up for my slow reflexes.

What if you had three eyes?

✧ Are there any examples of creatures that have three eyes?

✦ Yes—a number of fish, frogs, and lizards. In fact, people also have the beginnings of a third eye in their pineal gland buried in the middle of the brain. Like a true eye it reacts to light, but instead of transmitting images, it produces a hormone that affects our moods.

✧   Well, I'm not sure what value it is to react to light in the middle of the brain, but if I had a real third eye, I know where I'd want it to be.

✦   This isn't going to be something disgusting and perverted, is it?

✧   No. I just have a natural curiosity to know more about where I came from. . . .

✦   I knew it!

✧   I'd want the third eye to be in back of my head, so I could see where I'd been. That would also be a useful place for it to see if any predators or bill collectors were sneaking up on me.

✦   Actually, we already have some protection in that regard, given our senses of hearing and smell. Unlike vision, we hear sounds from all directions, whether or not our ears are directed at the source. Animals that do have to worry about predators sneaking up on them have a much better sense of hearing (and smell) than we do. So a third eye at the back of your head might not give you that much of an advantage compared to having much keener senses of hearing and smell.

✧   I just think the more eyes the better!

✦   Given how much of our brain is devoted to the processing of visual information, we might be better off having two really good eyes with more extensive information processing than having a third one.

✧   Well, if we did have a third one, where would you put it?

✦   At the end of a finger might be an interesting place. I could then extend my finger at arm's length, giving me about a three-foot distance between my third eye and my other two—about ten times more than my present eye separation.

✧   So what?

✦   With a much greater distance between my eyes, my parallax, or ability to judge distance, would be greatly improved. Also, by having the eye at the end of my finger, I could see behind me just as well as if I had the eye on the back of my head. Of course, there would be some serious drawbacks too.

✧   Like?

✦   One problem is that given the speed at which nerve impulses travel, you want to have your eyes as close to your brain as possible. After all, if a sharp twig is approaching, you don't want your brain to take all day to tell your eyelid to close. Also, the chances of hurting a finger are much more than those of hurting an eye located on your face.

✧   To say nothing of the problems of typing or playing the piano, or picking your nose.

✦   One solution might be to have the third eye located at the end of a retractable antenna located on top of your head. An even more interesting possibility might be to have part of the eye located at the end of the antenna.

✧ What do you mean part of an eye?

✦ The antenna might then be like a bundle of optical fibers that each transmit light directly to a "retina" located in the head—sort of like the fiber bundles they use for taking pictures inside the human body.

✧ Speaking about taking pictures inside the human body. . . .

What if everyone became blind?

✧ It is difficult to imagine how such a thing could happen—perhaps if there were some worldwide epidemic, or a space-borne virus.

✦ That's exactly the plot of the story "The Day of the Triffids," where aliens in the form of intelligent giant plants try to take over the world. Blinding everyone was the author's way of evening the odds in an invasion that would be difficult to take seriously otherwise.

✧ Triffids aside, it certainly would be a radically different world if everyone were blind. I wonder how much of today's civilization could continue to exist? Even though many blind people can function quite successfully in today's society, it would be an entirely different matter if everyone were blind. Let's consider some specific areas of technology. How about transportation, for example?

✦ It's difficult to imagine automobiles being modified to the point where blind people could drive them. On the other hand, considering that planes can be flown under conditions of near zero visibility, it's conceivable that they could be flown by blind pilots if instrument panels were modified—though making the necessary modifications after we were all blind would be quite tricky. Trains might be a more likely possibility, but in both cases the long term prospects appear remote.

✧ How so?

✦ Complex vehicles, such as trains and planes, must be maintained on a regular basis, and much of this must be done visually—inspecting planes for the development of hairline cracks, for example. Without such regular inspections, planes and trains might rapidly become completely unreliable.

✧ What about other areas, such as electric power production and building construction? Here also, it seems unlikely that complex power stations and buildings of the kind we now have could be built and maintained if everyone were blind.

✦ People might not care much if the lights went out, but our whole civilization runs on electricity, and without it we would be in trouble. As far as building construction goes, we certainly could not build structures of the kind we have today, but much simpler structures, such as tepees, could probably be built.

✧ Now that we've taken care of shelter, what about food? Could a blind civilization handle agriculture?

✦ Agriculture is currently a very high-technology operation, and farming is one of the most dangerous occupations, even for sighted people. It therefore seems highly unlikely that a blind farmer could produce crop yields close to the present ones. Most likely, we would have to revert to the days when a very significant fraction of the population was involved in growing its own food.

✧ It sounds like you're describing a reversion to a much lower level of technology than now exists.

✦ Yes, and with communication and transportation much more limited than what we now have, it seems likely that society would operate on a much more local basis than is now the case—much as it did before the industrial revolution. Quite possibly, any economy that existed would be strictly on a barter basis.

✧ What we're considering so far is everyone going blind today. But what if people had always been blind? Would they have been able to develop a civilization, especially if other creatures, some of them predators, were not blind?

✦ It seems highly dubious that even such "low" technologies like fire, metalworking, and agriculture could have been developed by blind humans. It also seems unlikely that people would be able to survive against predators much fiercer than Triffids if they were not blind.

✧ Of course, we've only considered the disadvantages of a society in which everyone was blind. Let's not forget the good side. Zoos could get by with having only elephants (which could pass for any animal), you could dress (or not) as you pleased, and no one would notice if you spilled ketchup on your shirt in a fancy restaurant. What do you think society would be like if only 99 percent of people went blind, rather than everyone?

✦ That would likely make a world of difference, in terms of being able to maintain today's technology. Sighted people would probably be regarded almost as gods. Their special power of sight would be a mystery to the rest of the population, and it would be comparable in today's world to someone who could reliably read minds.

What if you had
X-ray vision?

✧ Is X-ray vision really possible?

✦ Yes, but not using X rays. Firefighters use infrared cameras mounted on their helmets that allow them to see through smoke. Also, since the cameras respond to small temperature differences in different parts of a scene, they can let firefighters spot flames hidden behind walls or ceilings.

✧ If I had X-ray vision, I know what I'd use it for.

✦ I don't want to hear about your prurient fantasies. Anyway, if you are thinking about seeing through people's clothes, you can forget about it, because that wouldn't be possible.

✧ How come? Superman does it routinely, at least the part about looking through walls.

✦ Superman is a fictitious cartoon character; we are real cartoon characters—bound by the laws of physics. Superman's form of X-ray vision assumes that X rays emanate from his eyes like laser beams and can be made to penetrate to whatever depth he desires. But real eyes—be they of the optical or X-ray variety—are receivers, not emitters of radiation. If you had X-ray vision, surprisingly, you would not see anything at all unless the object you were looking at either emitted X rays of its own or was in the path of an X-ray beam.

✧ I assume the second method is how X-ray machines take pictures of the insides of our bodies.

✦ Exactly. X-ray pictures are, in effect, shadow pictures of objects that are partly opaque and partly transparent. As with shadow pictures of your hands on a wall, no lenses are needed with X-ray machines. But without using lenses, it is difficult to see how X rays could be focussed into an image on your retina—which is another reason to be skeptical about the possibility of X-ray vision.

✧ But suppose, unlike our actual eyes, I did have eyes (or some other organ) that did emit X rays. Couldn't I have a form of X-ray vision based on the amount of X rays that were reflected back from different layers of someone's skin and clothing?

✦ Back to seeing through people's clothes again, are we? The idea is still ridiculous, because the person's skin would have to be highly reflective of X rays, while the clothing would have to be virtually transparent. In the real world, both skin and clothing are transparent to X rays.

✧ But what about those magnetic resonance imaging (MRI) devices that are used for medical purposes? Can't they show internal cross sections of your body to any desired depth?

✦ Yes, but this would be like viewing a slice through a person taken at a fixed distance from your eye, which is not at all the same as seeing through clothing. Incidentally, an MRI-type of X-ray vision is about the only one that would be conceivable.

✧ How would it work?

✦ Much the way bats use ultrasonic echoes of their emitted clicks to form an image of their environment. Let's say you emitted some sort of signal that bounced off an object, penetrating it to some depth before it was reflected back. If you could separate the parts of the reflected signal that came from different parts of the object, you would, in effect, be viewing a slice through it.

✧ So how can I become a fictitious cartoon character, not bound by the laws of physics?

What if you could make yourself invisible at will?

✧ One way to achieve invisibility would be to use the method employed by The Shadow, the crime fighter who could supposedly "cloud men's minds."

✦ That would be a good way to do it, but in practice it seems difficult to imagine your being able to hypnotize anyone who just happened to be looking your way.

✧ Are there some approaches to invisibility that might be feasible in reality?

✦ Nature has endowed some of her creatures with such a high degree of camouflage that they are nearly invisible in their usual environment. A couple of examples include the chameleon and many kinds of fish and insects. Some chameleons and fish can even change color when the color beneath them changes, so as to retain their invisibility against the background. For example, a cuttlefish placed on a checkerboard can imitate the pattern on its body.

✧ Is there a way we could emulate this approach?

✦ In principle, yes. If we were being viewed from just one direction, we could imagine a cameralike device that would project an image on our front surface of the scene behind us that we were blocking. This approach works nicely for a fish that is always viewed from above, but it wouldn't work too well for us.

✧ Any other prospects for achieving invisibility?

✦ Certain solid materials such as glass, quartz, lucite, and diamond are transparent. If one of these materials is placed in a liquid having the same optical properties, it becomes invisible. Try looking at a glass rod, for example, in a glass of water. Even though the so-called refractive indices are slightly different, you may find that the submerged glass rod is nearly invisible.

✧ Could this method apply to making yourself invisible?

✦ Probably not, because it would not be enough to alter your body's surface features. You would have to have some way of altering your entire body's molecular structure to make it transparent to visible light, which seems somewhat far-fetched.

✧ Assuming it were possible for me to become invisible at will, I could go anywhere at any time without being observed. The possibilities are endless.

✦ I don't think I want to hear them. But another interesting possibility to think about would be if we were all invisible all the time. This, of course, would be different from everyone's being blind, because we would be able to see everything else—just not people.

✧ I always wondered about invisible people. If light just passes right through them without any effect, it seems as if they shouldn't be able to see anything. Seeing depends on light hitting the retina and being absorbed, which would not occur if you were invisible, so all invisible people should be blind.

✦ Aha; you have discovered the fallacy regarding complete invisibility.

✧ I guess I'd settle for making myself completely invisible except for my retina. Maybe The Shadow's method of clouding men's minds was the best solution after all. The Shadow knows what evil lurks in the hearts of men. . . .

What if you could move objects by thought alone?

✧ I do this every time I lift my arm up, but I assume you are talking about objects that are not part of my body. Haven't there been reports of this phenomenon—for example, people being able to bend spoons by thought alone?

✦ The supposed ability to move objects by thought alone, known as psychokinesis, or PK, has been claimed by a number of so-called psychics. Benders of spoons and keys never seem to be able to accomplish their wonders when strict controls are imposed preventing them from secretly doing the bending by physical means.

✧ But I thought these abilities were documented under the watchful eyes of at least one group of physicists at the Stanford Research Institute.

✦ Professional scientists are probably far poorer than professional magicians at imposing the tight controls necessary to ensure that no cheating takes place. Scientists used to working with electrons and gerbils don't expect them to cheat, but people sometimes do. When these tight controls are in place, the abilities of psychics seem to mysteriously disappear.

✧ But maybe the inability to demonstrate PK and other psychic powers under tight controls is a result of the delicate nature of the phenomenon, which is destroyed by the negative vibes of skeptical minds and controlled conditions.

✦ That's what some proponents of PK would have you believe. But until the phenomenon is observed to occur under specified repeatable conditions, it will never be accepted by scientists.

✧ But are there any ways psychokinesis could be real? Maybe my thought processes only influence something very slightly, so by thinking of the number six, I can make it come up on a roulette wheel a little more often than chance would predict.

✦ If you can, you're wasting your talents just talking about it. There are, in fact many serious, well-intentioned people trying to observe PK and other "psi" phenomena under controlled laboratory conditions. Many of these workers have reported positive results, but these studies all suffer from one or more serious problems.

✧ Such as?

✦ First, there's the "file drawer" problem: people reporting positive results in their studies are much more likely to publish their results than those finding negative results. Second, the statistical methods many studies use are highly

"Darn. I meant to think one lump, not two."

questionable. Third, it is one thing to say that many experiments give positive results and quite another to say that the results form a consistent pattern.

✧ Are there any PK workers who get consistent results over an extended time period?

✦ One of the most "successful" has been Robert Jahn, a former engineering dean at Princeton University, who had subjects using thought alone try to influence the output of a computer that was randomly generating coin flips. If the subjects had zero PK ability, you would expect that in several million trials someone thinking "heads" would find that the fraction of heads was 50 percent, assuming the random number generator was perfectly fair.

✧ What did Dr. Jahn find?

✦ He has claimed a positive result in his experiments, because 50.02 percent of the time his subjects obtain a head—the 0.02 percent above chance being supposedly a statistically significant result, given the large number of trials. But most of that "positive" result comes from just one of Jahn's six operators. Also, with positive effects so small, it is exceedingly difficult to rule out biases that could influence the results—for example, a slight bias in the random number generator.

✧ So is there any way in which PK might be real?

✦ Yes, as a matter of fact. Using a number of electrodes taped to your head, it is possible to detect and record your brain waves. These signals can be amplified and used to control external devices such as an electron beam producing a spot on a screen. Given the feedback of seeing the position of the spot, you should be able to control your brain waves so as to move the spot on the screen, or control motors. Apparently, the procedure has already been used to allow someone to steer a boat.

What if people could read minds?

✧ I know exactly what you are thinking about this subject: that mind reading isn't really possible. If I'm right, doesn't that prove there is something to the idea?

✦ You obviously based your guess on my previously expressed skepticism of other paranormal powers such as PK. Professed psychics often use exactly the same technique. Being good students of human nature, they can infer much about a person's beliefs or mental state from a bit of conversation.

✧ But my last conversation with my psychic hot-line friend, at just $4.95 a minute, described my mental state quite closely. How could they do it just by guessing?

✦ Most people can find ways to make sense of just about anything. As an experiment, a college class was provided with horoscopes that everyone thought described them extremely closely. It was then revealed that everyone was given the identical horoscope. This experiment showed that generalized statements about our personality or mental state easily can be interpreted to have a specificity that is unwarranted.

✧ But getting back to mind reading or telepathy, do you claim that it is scientifically disproven or impossible?

✦ Not at all—just that it has not yet been shown to exist. For us to document the existence of telepathy requires an experiment that can be replicated anywhere under specified conditions, so as to produce a well-defined positive result—which has never been done. In addition, exactly the same problems cited with studies of PK apply here as well: low reporting of negative results, inappropriate use of statistics, and inadequate controls against cheating. An extensive body of poorly done, often inconsistent experiments cannot make up for one conclusive repeatable one.

✧ But isn't it possible that telepathy is a rare phenomenon, showing up only at times of great emotional stress? For example, I have heard many stories of people realizing through a sudden revelation that a loved one had just died.

✦ It is not impossible that this could be due to telepathy. But coincidence might be a simpler explanation. We tend to forget the many times we might have such thoughts when nothing ever comes of them, but obviously we would remember for the rest of our lives the one occasion a loved one died about the same time we thought of it. On the other hand, if I myself experienced the thought and the event occurring at precisely the same time, I must confess, it would stretch the bounds of coincidence.

✧ Let's suppose that telepathy were real. How would it change our lives?

✦ If it were relatively rare, as in the case we just considered, it probably wouldn't change them at all. On the other hand, if we all could listen in to one another's thoughts on a continuous basis, no one could ever keep any secrets from anyone else. It would seem as if everyone would have a very clear picture of everyone else's intentions all the time, except for one little problem.

✧ What's that?

✦ The problem of crosstalk. The issue of their credibility aside, experiments done with telepathy do not show any falloff with distance in the strength of the "signal," as would be expected with physical phenomena such as radio signals. The experiments also don't show any shielding effect. The signal passes through all obstacles. If this were so, you would be receiving simultaneous messages of the same intensity from every living being (and—who knows?—maybe some dead ones too).

✧ Well, maybe you could only read the minds of people who were thinking about you and had tuned in to your frequency, or perhaps if you had tuned in to their frequency.

✦ I think I'm getting an unlisted number.

What if you had a premonition of a plane crash?

✧ I hope I got the premonition before I got onboard!

✦ Right. If you were already onboard and you had the nerve to tell the crew about it, you would likely be thought to be crazy or a terrorist.

✧ Well, I don't believe in premonitions, but if it were a really strong one, I might take a later flight if I could.

✦ Of course, your premonition of a crash might have been for the flight that you switched to! But let's assume that your premonition included a vision of the flight number. I probably would switch to a later flight too, especially since I also don't believe in premonitions.

✧ I don't get it. Why would you pay special attention to a premonition if you don't believe in them?

✦ I imagine that people who do believe in such things must get premonitions or visions often—some weak and some strong, and they would have a hard time sifting the true signals from the noise (if there is such a thing as a real premonition). But since I have never experienced such a feeling, I might be willing to take the first one more seriously.

✧ Why just the first one?

✦ Well, if the first one turned out to be a false alarm, I would view future premonitions with skepticism.

✧ Suppose the first premonition turned out to be true?

✦ Seeing the future can be a tricky business. Most "seers" make their predictions of the future in very general terms. If my premonition was just of a plane crash without any more specifics, I probably wouldn't be too impressed with my abilities.

✧ But I have heard about many people who have had very specific premonitions about disasters or deaths. Isn't the probability of this happening by chance virtually impossible?

✦ It is extremely difficult to evaluate such probabilities, because you would need to know how often a person gets premonitions and how often those premonitions turn out to be false. Often we may get fleeting thoughts of death or disaster that don't fully register in our consciousness. But they would certainly be remembered later if something were actually to happen. The ability of "seers" to predict the future is quite unimpressive when you include all the false predictions.

✧ OK, but let's say you had a very precise premonition of a plane crash, and it was the first one you ever had.

✦ If the first premonition of my life was of a crash of a particular flight that I decided not to board as a result of the premonition and the plane actually crashed, then I would become a believer in the ability to foresee the future. In fact, I would probably spend the rest of my life trying to understand the phenomenon.

✧ Heavy, man. Suppose you got a premonition about a specific flight crashing that you weren't going to take?

✦ If no friends or loved ones were scheduled to take the flight, I don't think I'd have the nerve to alert the authorities on the strength of my first premonition.

✧ I take it that should you get a future premonition after your one correct one, you would alert the authorities.

✦ At the risk of being thought a crank I would. But when I got my first premonition, I would be sure to tell as many people as possible about it.

✧ To establish your credentials as a seer if a crash occurred?

✦ Yes, and to be reminded of my foolishness if it didn't.

What if humans had a dog's sense of smell?

✧ I know smell is not one of our keener senses—a fact I'm often quite happy about when I'm around my friend Stinky. Incidentally, why don't people seem to notice their own body odor, even when it's obvious to everyone else?

✦ Because smell, like our other senses, becomes "fatigued" if exposed to a stimulus for a prolonged period of time. Being around your own body odor continuously, you tend not to notice it after a while, even though you still can detect other odors just as well. The same phenomenon would happen if you were around a steady sound continuously.

✧ How sensitive is a dog's sense of smell?

✦ Dogs have a sense of smell that is roughly a million times better than ours, particularly in their ability to detect skin secretions of other mammals. This is how bloodhounds, for example, are able to follow trails that are hours old.

✧ What exactly is the bloodhound detecting when it smells an odor?

✦ A person or animal's body odor consists of a large number of individual molecules that create a specific identifiable smell sensation when they reach the nose.

✧ How does the nose recognize one particular smell?

✦ The individual components of a complex molecule have specific shapes that fit into particular receptor cells just like a key fits a lock. A dog's nose essentially does a breakdown of the individual molecular components, in a manner analogous to the way a chemist identifies an unknown molecule. In fact, a smell such as that of underarm sweat contains over 250 components, and the resulting combination, which is unique to a particular person, constitutes his or her "odorgram."

✧   But what does the trail consist of that a bloodhound follows?

✦   The molecules shed by our skin leave a trail on the ground that remains for a considerable time until it is gradually dispersed by the wind. Unlike humans, who have a fairly narrow repertoire of basic smells that can be recognized, dogs appear to be extremely discriminating in being able to sense small differences in these smells.

✧   When I think of some of the things my dog smells, the last word that comes to mind to describe his nose is "discriminating." Are there cases of an even keener sense of smell in nature?

✦   The record is held by the male silkworm, who is able to detect an interested female at a distance of 2 miles, based on her secretions. Experiments show that many other insects apparently have a similar sensitivity. But unlike dogs and other mammals, an insect's sense of smell is extremely specific to one particular odor— that emitted by a potential mate.

✧   Somehow, I can't picture scientists interviewing silkworms to learn how strongly they perceive different odors.

✦   What they actually do is to connect a meter to the output from the insect's smell receptor. They observe that the instrument doesn't budge for odors other than that of an eager female, and this odor causes the meter to shoot off scale.

✧   Judging from the persistence of neighborhood dogs when my dog goes into heat, apparently sex and smell must have an important connection in mammals too.

✦   Humans also may make a connection between sex and smell, which may have been more explicit in less "civilized" times. At least that's one possible reason evolution has favored the retention of odor-enhancing underarm and genital hair, even while we have shed most of the rest. Perhaps at some primitive level we find one another's smells appealing despite our abundant use of deodorants to hide them.

✧   I'll explain that to my girlfriend. What would life be like if we had a dog's sense of smell?

✦   Besides the connection with sex, dogs and other animals rely on their sense of smell to search for food, recognize other animals, and sense danger. Given its acuity, dogs rely on their sense of smell much more than their sense of sight. Undoubtedly, if we had such an acute sense of smell, we would have a much richer vocabulary for talking about smells. It would give us a whole new perspective on our surroundings and might even allow us to make observations and judgments about other people's emotional states, and their attitude toward us, based on their smell.

Chapter 7

Disasters

What if humans became extinct?

✧ A natural question is, what would be cause of our demise? For a while, most people suggested nuclear war, but now that the nuclear threat has been lifted, the betting is probably on some environmental catastrophe.

✦ But you shouldn't think the nuclear threat has disappeared, just because the Soviet Union is gone. The nukes are still there, and political developments or widespread weapons proliferation could bring the threat of nuclear annihilation back with a vengeance.

✧ That's a cheery thought. So I take it you believe that nuclear war would be the most likely cause of human extinction.

✦ Actually, no. As horrific as a large-scale nuclear war is likely to be, its consequences would probably stop short of human extinction. The straggling survivors might be hard put to reconstruct a society anything like what now exists, but some areas of the globe would probably be relatively untouched by the short-term effects of the war. The potentially serious long-term effects, which would be worldwide in scope, are also unlikely to bring about our ultimate demise.

✧ So if nuclear war doesn't do us in, what might?

✦ A roundup of the usual suspects would include overpopulation, pollution, a runaway greenhouse effect, a large asteroid impact, and a super AIDS-like epidemic, or maybe some combination of these.

✧ Is there any significance to the order in which you listed these unpleasant possibilities?

✦ Yes, based on my own completely subjective feeling as to how likely they are to lead to our extinction—the order being one of increasing likelihood. I've listed overpopulation first only because it is unlikely to lead to our extinction by itself. Of course, overpopulation is one of the root causes of some of the other items, so it is extremely important to worry about.

✧ Are there any other possible causes of our extinction not included among the usual suspects?

✦ Yes, I would suggest two more. One that people probably used to worry about more than now is an invasion by aliens. Another one, also popularized in various science fiction movies, is our developing robots with supercomputer brains who decide that they have no further use for us or that we represent a threat to the well-being of the planet. In my opinion, even though the previously listed threats could make life very unpleasant for humans, this last one has the greatest chance of leading to our extinction.

✧ You can't be serious. Why couldn't we design the robots so that they would never get such ideas?

✦ In principle, the robots could be programmed so that avoiding harm to humans was their highest objective—the First Law of Robotics in Isaac Asimov's science fiction stories. On the other hand, it is impossible to be sure that a complex technology will always work as planned. One scenario that has been suggested is the possibility of designing self-reproducing robots that could undergo evolution just like living creatures. In such a scenario we cannot rule out the possibility of a "mutant" robot who didn't obey the First Law of Robotics, giving rise to new robots who had no use for humans.

✧ Hmnn. Maybe I'd better think twice about the new memory I was going to get for my computer. Do you think we will go the way of the dinosaurs any time soon?

✦ The dinosaurs, despite their eventual extinction, were tremendously successful animals: they lived on this planet for around 150 million years—60 times as long as we have so far. Judging from the accelerating pace of change taking place today and all the extinction threats we have looked at, I'd bet on the dinosaurs' record time for "king of the hill" status beating ours.

✧ If we do become extinct, and it's not aliens or robots that follow us, my bet would be on the cockroaches.

What if we had another ice age?

✧ I thought global warming was the thing we needed to worry about now.

✦ Toss in a string of colder than normal years, and for sure we'll be more concerned about ice ages than global warming.

✧ But don't we have computer models that can forecast Earth's climate?

✦ We have some idea of how a given specific change in the atmosphere will change the long-term climate, but we really don't have a clue what determines the underlying long-term variations in climate. Assuming that it's been getting warmer in recent years, there is no way to know whether the warming is due to the usual underlying natural variations or a relatively new greenhouse effect—whether natural or artificial. In fact, for all we know, we may be in the middle of an ice age right now.

✧ How can that be?

✦ Previous ice ages typically have lasted many millions of years. Within each ice age there were warm spells lasting tens or hundreds of thousands of years and occurring at irregular intervals. We could be in the middle of one of those warm spells.

✧ How long has it been since the last cold spell?

✦ Aside from short "mini-ice ages," the last one was about ten thousand years ago. Geologists can track the advancing and retreating ice sheets because of the erosion they produced on rock surfaces. Ten thousand years ago, ice sheets covered much of Europe and North America, advancing as far south as New York City.

✧ With all that water "locked up" in ice, I imagine there would have been quite a drop in sea level.

✦ Yes. In fact, during the last glacial advance sea level probably dropped about 100 meters or 330 feet.

✧ How about the opposite scenario? What would happen to the sea level if the present polar ice caps were to melt?

✦ In that case, the sea level would rise by about 50 meters, and most of the world's coastal cities would be flooded.

✧ It sounds as though New York is in big trouble whichever way the climate goes. I assume it would be bad news for life on Earth if we had another ice age.

✦ The geological record does show that ice ages bring about a drastic redistribution of plant and animal life on the planet, with many species becoming extinct. A particularly severe ice age occurred about 600 million years ago, before animal life proliferated across the planet. That ice age probably restricted marine life to a few habitable areas near the equator.

✧ Didn't you say that the cause of ice ages is not really known?

✦ That's right. Anything that can bring about a slight cooling lasting for a long time would be a candidate. An increase of volcanic activity or a large asteroid impact would both throw a large amount of dust and other aerosols into the atmosphere. In either case, the blockage of sunlight might cause enough of a temperature drop—it would take perhaps only a few degrees—to bring about another ice age.

✧ How could such a small temperature drop bring about an ice age?

✦ Through the mechanism of positive feedback, meaning that a small change "snowballs"—a particularly apt analogy. For example, if you were to cool the planet a bit, causing the ice sheets to expand slightly, the extra whiteness would lead to less absorption of solar radiation. It's exactly like the difference in warmth you feel when wearing light versus dark clothing on a sunny day. The expanded ice sheets would cause the planet to cool still more, causing the ice sheets to expand further.

✧   . . . And the next thing you know, New Yorkers are ice skating in summer on the Hudson River—assuming we don't become extinct!

What if there were an earthquake every day?

✧   I think I'd move to San Francisco. As long as everyone was worrying about earthquakes, no matter where they lived, you might as well live in a nice climate. How many earthquakes are there per year now, by the way?

✦   People often feel small earthquakes that cause no destruction—in Tokyo you'd feel fifty per year. In fact, the great majority of all earthquakes are so small they can be detected only by sensitive instruments. Thousands of these microtremors occur daily. Even though larger quakes are relatively rare, potentially quite destructive quakes of magnitude 6 on the Richter scale occur around a hundred times per year.

✧   What is that Richter-scale business all about?

✦   A given number on the Richter scale corresponds to the amount of energy released in a quake. For example, a major quake (7 on the Richter scale) would release an amount of energy equivalent to about 500 Hiroshima bombs. A one-unit change in Richter magnitude corresponds to a factor of about 30 increase or decrease in energy. So a devastating magnitude 8 earthquake, of which one or two occur per year, would be equivalent to about 15,000 Hiroshima bombs.

✧   Is 10 the maximum on the Richter scale?

✦   No, there is no maximum, since the scale measures the energy released. The 1906 San Francisco quake (magnitude 8.3) was one of the biggest ever recorded. Generally, the bigger the quake, the less frequently another one like it occurs. So in a typical year, there are a million microtremors, 10,000 minor shocks, 1,000 damaging shocks, 100 destructive shocks, 10 major earthquakes, and. . . .

✧   . . . a partridge in a pear tree. Do we have any way of predicting when an earthquake will occur?

✦   Many people have come up with schemes to predict earthquakes, ranging from observing strange behavior in animals to detecting certain changes in the weather prior to an earthquake, but as yet we have no reliable predictor of a major quake, except that we do know that once a quake occurs, aftershocks are likely to follow.

✧ How about the location of earthquakes?

✦ Earthquakes mostly occur at places where the twenty-some plates comprising Earth's crust border each other. The motion of these crustal plates is driven by Earth's internal heat. As these plates try to slide past each other like giant pieces of a jigsaw puzzle, they sometimes stick and then slip, with a tremendous release of energy. One of those plate boundaries runs down the West Coasts of North and South America.

✧ Which explains why the San Francisco area has had a number of quakes. But didn't the biggest quake ever to hit the United States occur in the Midwest?

✦ That's right. Three huge quakes hit New Madrid, Missouri, in 1811–1812. The Northeast, which is usually free of earthquakes, has had some big ones too. Perhaps the crustal plates occasionally get hung up at places other than their edges.

✧ How about the damage that a major earthquake can inflict?

✦ An earthquake hitting a major city can inflict a devastating level of damage, especially if the buildings are not constructed to be earthquake resistant. If a major quake were to hit a densely populated city such as Tokyo, rupturing gas lines and starting fires, the number killed in a firestorm following the quake might far exceed the toll from the quake itself.

✧ What would life be like if a major earthquake occurred every day?

✦ Right now, a major quake occurs somewhere in the world every month, so we are assuming quakes would occur thirty times more often. Although a catastrophic earthquake can lead to a huge loss of life, the fraction of all people killed by earthquakes would still be minuscule, given the tiny fraction that now die as a result of them. But the extent of property damage would become a major drain on society.

✧ How would we cope with the situation?

✦ We probably would drastically alter construction techniques. If we continued to build skyscrapers, we would want to put them on base isolators—now installed under some Japanese skyscrapers. These devices allow the most damaging back-and-forth waves of an earthquake to shake the ground underneath a building while it remains stationary. In some parts of the world, people would just become accustomed to earth tremors occurring on a daily basis.

✧ How come the back-and-forth shakes are the worst? I would have thought the up-and-down shakes would be just as bad.

✦ Put a salt or pepper shaker (representing a building) on top of this book (representing the ground) and start shaking the book. You will probably find that back-and-forth shakes are more likely to topple the "building" than up-and-down shakes.

✧ I guess if the earth were shaking all the time, we'd be in much the same position as birds living in wind-shaken trees—but without the benefit of wings!

Marriages between seismologists are shaky because they are always finding faults.

What if an asteroid struck Earth?

✧ Don't tell me the sky really is falling.

✦ Earth is continually being struck by chunks of rock and metal known as meteorites. The small ones burn up in the atmosphere before they hit the ground, but the ones we need to worry about are the larger, much rarer ones that survive that fiery trip through the atmosphere. Asteroids are what we call the really big guys.

✧ Where do all these asteroids and meteorites come from?

✦ They're part of the debris that didn't coalesce into planets when the solar system formed. Most of the asteroids orbit the Sun—especially in the asteroid belt between Mars and Jupiter, but some of them are in orbits that cross Earth's orbit.

✧ I imagine that if a big asteroid hit Earth, it would create a large crater. Doesn't the absence of such craters show there haven't been many big impacts?

✦ Actually, well over a hundred large impact craters have been found. They are usually somewhat difficult to spot because of weathering and erosion. But just take a look at the features on the face of our moon if you want to see a permanent record of crater impacts through Earth's history.

✧ So what are the chances of an asteroid impact?

✦ It depends entirely on the size of the asteroid. Suppose we consider an asteroid a mile or more in diameter, which might do to us what one apparently did to the dinosaurs 65 million years ago. Based on estimates of the number of such asteroids in Earth-crossing orbits, the chances of an impact with Earth during your lifetime are around 1 in 20,000—about the same as your chances of dying in a plane crash.

✧ I've heard that such an asteroid would impact with an energy release equivalent to 10,000 hydrogen bombs. It's difficult to imagine anything on Earth surviving that.

✦ Actually, despite its enormous explosive force, the impact, being relatively localized, would cause little worldwide destruction. The main problem would be that a sizable fraction of the asteroid would vaporize on impact and the dust would be thrown up into the atmosphere, where it would block off sunlight for many months. Earth's climate could be profoundly altered—possibly to the point where many species would become extinct.

✧ But how could the vaporized dust block off sunlight across the planet? Wouldn't it just be thrown up in the vicinity of the impact?

✦ The impact would be so violent that much of the dust would be thrown as high as the stratosphere, where there is no weather, so it would stay up there a long time and gradually spread around the whole globe before coming down.

✧ Couldn't we just shoot a missile to blow up any asteroid that we saw approaching Earth?

✦ By the time an asteroid on a collision course with Earth was spotted, it would be too late to do a thing about it. But fortunately, asteroids follow regular orbits, so with a good tracking system in place, we might have twenty or thirty years to do something about it—such as launching a nuclear missile to change its orbit very slightly. That's the good news.

✧ OK, what's the bad news?

✦ Asteroids aren't the only space threat. Comets, which follow highly eccentric orbits around the Sun, might not be spotted until only a year or two before impact. We might not be able to deflect them on such short notice, and trying to blow them up might just make things worse, causing many smaller pieces to strike Earth. The 1994 collision of a comet with Jupiter was a preview of what could happen if one struck Earth.

✧ I think I'll check my insurance policy to see if I have collision insurance. Actually, come to think of it, a better idea might be to start my own asteroid insurance company—who'd be around to collect?

What if a runaway greenhouse effect occurs?

✧ I don't think I'm going to like this one. Greenhouse effects cause unpleasant changes in our climate, right?

✦ Not necessarily. Without nature's own greenhouse effect, the average temperature of the planet would be a chilly zero degrees Fahrenheit.

✧ OK, so how does the greenhouse effect work?

✦ There are many gases in the atmosphere that are transparent to light but not to infrared radiation. These "greenhouse gases" therefore allow in the sun's warming rays but trap the heat (in the form of infrared) that is radiated from the warmed Earth. The glass in a real greenhouse, or in a car with sealed windows, does the same thing on a hot day since glass is transparent to light but opaque to heat radiation.

✧ What are the sources of greenhouse gases?

✦ By far the biggest source is water vapor, but there are many others, including methane—a product of paddy fields, termites, and bovine flatulence, which adds 35 million tons of methane to the atmosphere each year.

✧ Thirty-five million tons! What are those cows eating? But if the main cause of the greenhouse effect is water vapor in the atmosphere put there by nature, what's there to worry about?

✦ Due to human activities—mainly burning fossil fuels and clearing forests—the amount of carbon dioxide (CO_2) in the atmosphere (now less than 10 percent of the amount of water vapor) is increasing. Apparently, the amount of methane is increasing even faster, so that by the year 2050 its contribution will outweigh that due to carbon dioxide. Estimates are that if the atmospheric CO_2 and methane were to double, the worldwide temperature could rise by anywhere from 3 to 8 degrees Fahrenheit. But there are lots of uncertainties.

✧ So why don't we just wait and see what happens, and take some drastic measures only if we see an actual temperature rise?

✦ The problem is the long time delay between the addition of greenhouse gases and the climate change they cause decades later. One reason for the delay is the positive feedback that a small change can cause: add some CO_2 to the atmosphere and you cause some global warming, which evaporates more water vapor, which causes more global warming later, et cetera, et cetera. Given the delays, by the time the total warming trend is clear, it might be too late to do anything about it.

✧ Well, a temperature rise of 3 to 8 degrees Fahrenheit doesn't sound too bad. People in many parts of the world might be delighted with such a change.

✦ There would be "winners" and "losers" if the average worldwide temperature changed by as much as 8 degrees, but the losers would probably far outnumber the winners—particularly if the polar ice caps were to melt and flood all the world's coastal cities.

✧ What's the worst-case scenario if we do nothing about the problem?

✦ We could conceivably wind up with a runaway greenhouse effect, where positive feedback loops keep driving the temperature upward as the insulating blanket of water vapor gets thicker and thicker. The planet Venus probably experienced such a runaway greenhouse effect in its early history.

✧ What did the Venusians do that caused such a catastrophe?

✦ Given conditions on Venus, it's highly unlikely any Venusians ever existed. The runaway greenhouse effect on Venus was due to its proximity to the Sun, which results in the planet receiving about twice the solar energy that we do. In Venus's early history, the solar heating either prevented oceans from forming or else evaporated them, giving rise to a thick blanket of water vapor. As volcanos added CO_2 to the atmosphere, more and more of it accumulated, since there were no oceans to absorb it.

✧ So how bad is the atmosphere on Venus now?

✦ It is probably the most inhospitable planet in the solar system. The atmosphere is 100 times denser than ours, and its temperature is 880 degrees Fahrenheit—hot enough to melt lead. And by the way, if they existed, they'd probably be called Venereans.

✧   Well, I guess that's still another reason not to visit the place.

What if antibiotics stopped working?

✧   They could collect unemployment insurance? OK, what are antibiotics, exactly, and how do they work in the first place?

✦   Our knowledge of the curative powers of molds and other natural substances can be traced back to the ancient Egyptians. But the first antibiotic resulted from a chance observation by Alexander Fleming in 1928 that the fungus in bread mold killed a common bacterium, staphylococcus aureus, that we'll call by its nickname, staph. Fleming isolated a substance from the fungus, penicillin, that was the first in a series of antibiotics that scientists developed, mainly in the 1940s and 1950s.

✧   How do antibiotics kill bad guys like staph?

✦   Antibiotics are substances produced by small cells or microbes that can kill other cells. The trick is to identify the very small percentage of antibiotics that will kill germs without having an effect on the cells of the human body.

✧   Those antibiotics must be pretty smart. How do they do that?

✦   Bacterial cells are structurally different from human cells. For example, they usually have a protective outer coat, and they contain other structures absent in human cells. By interfering with the formation of the cell wall, penicillin, for example, can selectively destroy bacteria without harming the cells of the body.

✧   Why should antibiotics stop working if they have been so effective in killing the bad guys so far?

✦   Among any given type of bacteria, mutations can lead to rare strains that have a greater resistance to the antibiotic. These mutant strains survive in much greater numbers than the others, and they pass along their antibiotic resistance to their descendants. Fortunately, for us, most bacteria that have developed resistance to one antibiotic can be killed by another, but some of them—the so-called superbugs—have been developing resistance to a broad range of antibiotics.

✧   How did we get into this mess?

✦   Mostly, we have ourselves to blame. Have you ever stopped taking a prescription before the dose was gone, because you felt better? If you had a serious infection like tuberculosis, that seemingly harmless act could lead to the development of drug-resistant strains of bacteria in your body as the bacterial survivors of the medication regrouped and multiplied. We also accelerate the development of drug-resistant strains of bacteria by overusing antibiotics to the extent we do—especially in livestock, which routinely receive thirty times the human dose. But even without overuse, the problem would have occurred anyway, though not as soon.

✧ How bad has the problem become?

✦ A few years ago, very few of the 25 million annual ear infections among U.S. children were caused by a superbug, but soon all will be. Many hospitals have now become breeding grounds for the superbugs, and thousands of patients die each year from epidemics in hospitals.

✧ And I thought hospitals were a place to go to get well! How bad could it get if the superbugs achieve resistance to all antibiotics?

✦ We would be faced with a public health nightmare, with millions of untreatable infections. Mortality rates would climb to those of the 1930s or worse, where one in five people used to die of such diseases as pneumonia and tuberculosis.

✧ Presumably, scientists are busy in their laboratories working to find new antibiotics to stop the superbugs?

✦ Right. At the same time, the superbugs are busily developing into superduperbugs to overcome the new antibiotics! But it's not as if germs are out to kill us. They are just trying to survive, and for some of them we happen to be their food. Of course other, "good" bacteria inside our intestines are absolutely necessary in order for us to digest our food. In fact, without bacteria, we along with all other higher forms of life couldn't decay and have our atoms recycled into future living organisms—life as we know it could not exist.

What if a super- AIDS epidemic occurred?

✧ How can anything be worse than AIDS, which is nearly always fatal?

✦ As bad as the AIDS epidemic is, there are diseases that could lead to even more horrible consequences. For one thing, someone infected with the HIV virus that causes AIDS can have many symptom-free years without coming down with the disease. For another, AIDS, which is transmitted during sex and through blood products, is not so easy to catch, especially if you take "safe sex" precautions.

✧ I've just started a local chapter of "Sex without Partners."

✦ That might be the safest form of sex, but the use of a condom can reduce the chances of transmitting the HIV virus by about 90 percent.

✧   What are some diseases that are worse than AIDS?

◆   A particularly nasty one is Ebola, named after the Ebola River in Zaire, where the first outbreak was discovered in 1976. Like AIDS, Ebola is nearly always fatal, and it appears to be transmitted through bodily fluids, although originally scientists feared you could catch it just by breathing air in the presence of an infected victim. Ebola completely wiped out a dozen African villages during its 1976 outbreak, and it surfaced again in 1989 and 1995.

✧   That sounds scary. Where did Ebola come from, and how come it first surfaced in 1976?

◆   Like AIDS, Ebola is believed to have originated as a virus that only infected monkeys. Then somehow it changed slightly and became capable of infecting people. One major Ebola scare in the United States occurred in 1989, when a shipment of monkeys was found to be infected with Ebola.

✧   What happened?

◆   Even though the virus was transmitted to some of the scientists handling the monkeys, none of them came down with the disease. Apparently, there are subtle differences between the version of Ebola that is fatal to monkeys and the version that is fatal to people.

✧   What would happen if a disease like the human version of Ebola ever got a foothold on U.S. shores?

◆   If the virus ever mutated to a form that could be transmitted through the air, the results could be devastating. An easily transmitted, rapidly fatal disease would create a catastrophe in a relatively mobile society such as ours. The loss of life from a large-scale epidemic could be comparable to the aftermath of a nuclear war.

✧   Could anything be done if this happened?

◆   Remember that viruses, unlike bacteria, cannot be killed by antibiotics, and the only effective treatment currently is a vaccine that prevents infection in the first place. Very likely it would not be possible to develop a vaccine in time to prevent the virus's large-scale spread, considering that even after many years we have no vaccine against AIDS.

✧   Is there any likelihood of this actually happening?

◆   New strains of viruses occur all the time. Because we don't understand exactly how viruses make the jump from animal varieties to the human variety, we need to be continually alert to new viruses that arise anywhere in the world. In the past, when people were much less mobile, a new lethal virus might not spread widely. But in today's world of frequent air travel, a lethal virus arising anywhere could rapidly spread worldwide.

✧   Especially if scientists go to study it and bring it back!

What if two small countries had a nuclear war?

✧   I guess a lot would depend on whether the countries involved had just a few small nukes or something more extensive. Incidentally, what is the nuclear status of various countries?

✦   The list of countries now in the "nuclear club" currently includes the United States, Russia, Kazakhstan, Ukraine, England, France, China, India, Pakistan, Israel, and possibly South Africa and North Korea. Although the last five countries have not admitted to having a nuclear arsenal, most observers include them on the list. A large number of other countries have the ability to build a large nuclear arsenal anytime they choose to.

✧   Do all the nuclear countries have a large number of nuclear weapons?

✦   The United States and Russia have the most nukes—in the tens of thousands. It seems likely that any country that opted to go nuclear would choose to build a militarily significant number rather than a token arsenal simply to proclaim their membership in the nuclear club. Of course, some of the nuclear countries may not have finished building their arsenal and may not yet have a large number of weapons.

✧   The conventional wisdom on nuclear wars used to be that they could not stay limited once they started. Do you think that continues to be true?

✦   Not necessarily. Much would depend on which countries were involved. A war involving some of the countries in the nuclear club would likely drag other countries including the United States into the conflict. On the other hand, the likelihood of this happening is probably less now in the post–Cold War world than in an earlier era. Once the nukes started flying, some countries might well rethink their peacetime pledges and commitments.

✧   How about the physical effects of a "small" nuclear war, especially radioactive fallout; wouldn't it inevitably spread to uninvolved countries?

✦   Radioactive fallout would be dispersed around the world, but during the months and years this dispersal would take, the fallout would decay to fairly low levels. Except for countries directly downwind from the conflict, the fallout levels would, at worst, cause a statistically very small increase in cancer rates.

✧   But what about the nuclear-plant explosion at Chernobyl, where radioactive fallout contaminated milk in countries remote from the site of the disaster?

✦   "Contamination" is always a matter of degree. You could say that our bodies are contaminated all the time, if we count the small amount of radioactivity always present. The real issue is how much harm is done by a particular level of contamination, and how concerned we should be about it. We normally set extremely stringent limits on radiation exposure, so that we would regard milk as contaminated even if it contained minute amounts of fallout whose actual health effects would be too small to measure. On the other hand, even with the much larger contamination caused by a nuclear war, the likely increases in cancer rates would still be very small in countries not directly downwind of detonations.

✧ How about other effects of a small nuclear war?

✦ The greatest worldwide effect might be the psychological one. Don't forget that nuclear weapons have not been used in warfare since the United States dropped two on Japan at the conclusion of World War II. During much of the nuclear age, the conventional wisdom has been that nuclear weapons play an important role in deterring war, and in particular deterring other countries from using their nuclear weapons, but that their actual use in warfare would be a disaster.

✧ What might the psychological impact of a small nuclear war be?

✦ Three very different trends are possible. You can imagine that most countries would become so repelled by the awesome devastation and death that the long-sought goal of worldwide control of nuclear weapons would be given new impetus, and real steps taken toward nuclear disarmament. Another possibility would be that many countries, becoming suddenly aware of their own vulnerability, would seek to build defenses against nuclear weapons delivered by ballistic missiles.

✧ What's the third possibility?

✦ The third possibility depends on just how the small nuclear war turned out to be. Conventional wisdom says that in any nuclear war both sides will inevitably be losers. If, in fact, one country should wind up being a much bigger loser than the other, those nonnuclear countries capable of building the bomb might overcome their nuclear aversion in a hurry. A small nuclear war might well set off a worldwide arms race.

✧ I just thought of a fourth depressing possibility. We in the West have become so saturated with stories of suffering around the world that a small nuclear war might simply pass by as one more horrible event among many. In fact, in the aftermath of a small nuclear war, the radiation might make it unfeasible for television crews to bring on-the-scene gory pictures into our living rooms. As a result, the psychological impact might even be less than that of other past tragedies.

What if all nuclear weapons were destroyed?

✧ "No nukes is good nukes." That would be wonderful—we would no longer be haunted by the possibility of nuclear war.

✦   It would, of course, not be possible to destroy the knowledge of how they are made, so the possibility of reconstructing large nuclear arsenals would remain forever.

✧   We could, perhaps, burn all the physics books and kill all nuclear scientists in the interests of achieving a nonnuclear future, but that might be a bit extreme.

✦   In an age when even a bright undergraduate physics student can design a crude nuclear bomb, your idea probably wouldn't work anyhow.[1]

✧   But even if the possibility of reconstructing nuclear weapons remained forever, wouldn't it be a blessing if all nukes were destroyed?

✦   That's far from clear. A good argument can be made that the large arsenals developed since World War II are precisely what kept the United States and the former Soviet Union from direct confrontation during the Cold War. With the knowledge that any war could easily escalate to a nuclear exchange, both sides were perhaps especially careful to avoid direct conflict.

✧   But wouldn't the mere existence of large arsenals increase the likelihood of an accidental nuclear war, or an escalation from a conventional to a nuclear war? Given the damage that nuclear weapons can inflict, wouldn't each side want to use its nukes before they were destroyed by the other side's first strike?

✦   As long as both sides had an invulnerable large force of nukes in their submarines on continuous patrol, they could remain confident that they would have enough nuclear firepower left after the other side attacked to inflict a devastating retaliation.

✧   OK, granted that large arsenals might have been justified during the Cold War, wouldn't everyone be better off now if we destroyed all the nukes, or at least drastically reduced their number to a "symbolic" level?

✦   The problem with a very small arsenal is precisely that it might make a much bigger difference who "shoots" first, particularly if the initial attack wiped out the government and military communications of the other side.

✧   OK, let's assume we get a worldwide agreement to destroy every last nuke.

✦   Verification of such an agreement would probably pose political and technical challenges that would be impossible to overcome. Using remote detectors (on satellites), it would not be possible to find out if another country had a well-hidden cache of weapons.

✧   But for the sake of argument, suppose it were possible to find any hidden

---

1. Fortunately, a bomb built by most undergraduate physics students probably wouldn't work either, because even if a student could get the necessary materials, a considerable gap exists between a design on paper and a working weapon. But the bad news is that other ways to kill large numbers of people do not require a very high degree of technical sophistication.

arsenals with confidence. Wouldn't it be a good thing to destroy all nukes in that case?

✦ Presumably, nations would still possess industrial facilities that could be pressed into reconstructing nukes during a period of heightened tension. If one nation felt that it could assemble a large arsenal on very short notice before its rival did, it might be tempted to use its temporary nuclear advantage. That temptation would be much greater than it is now when both countries have lots of nukes.

✧ Surely the situation cannot be as hopeless as you describe. There has to be some way for the world to put the nuclear genie back in the bottle.

✦ The only possibility that has a chance of working would be to have all nuclear weapons under international control by a body that had much stronger "teeth" than the United Nations. Whether this would be a positive or negative political development for the world would depend on the intentions of the people controlling this supranational body. Nuclear weapons in the hands of a central government could either enslave the world or liberate it from the threat of nuclear war.

✧ Or maybe both? Too bad it's not feasible to develop "antinuclear bombs" to prevent nuclear bombs from detonating!

What if there were a large nuclear war?

✧ I've read that a large nuclear war could mean the extinction of humanity, and possibly all life on Earth.

✦ The key word in your comment is *could*. As horrible as the consequences of a large-scale nuclear war are likely to be, most studies don't indicate that these most dire consequences would occur. On the other hand, the effects might be worse than our best guess.

✧ How can we even estimate what the effects of a large-scale nuclear war might be, when such a thing has never occurred?

✦ Virtually all the estimates require a range of assumptions that at best are simply educated guesses, and the range of uncertainty is usually quite large. We know what the blast effect of a weapon of given size is at any given distance, but to say how many people would be killed or wounded, or how likely a firestorm would be, is not easy. Some people believe that in making these kinds of estimates, we should always err on the conservative ("worst case") side, so that we never fall into the belief that a nuclear war could be fought and won.

✧ Don't they have a good point?

✦ Exaggeration of the effects of a nuclear war, even with good motivation, is probably unwise, because it leads to the perception that the science is being driven by politics. It is best to be up-front about all the uncertainties inherent in such estimates.

✧   What effects would kill the most people in a large nuclear war?

✦   Most people in or near population centers or military targets would probably be killed by the direct blast of the weapons or the fires they might ignite. But neither of these effects would cause devastation worldwide, at least in the short term.

✧   How about the long term? What are some worldwide effects of a large-scale nuclear war that have the potential of doing the greatest damage to the planet as a whole?

✦   The three global effects that have received the most attention are radioactive fallout, ozone layer depletion, and the climatic change that has been referred to as "nuclear winter."

For commanders in chief, the combination of uncertainty and dyslexia can be catastrophic.

◇ What might the impact of these effects be?

✦ A layer of ozone in the upper atmosphere protects us from ultraviolet solar radiation, but ozone layer depletion would probably be the least serious of the three global effects. Although a large-scale nuclear war could destroy as much as 80 percent of the ozone layer, natural processes would probably restore most of it in just a few years. During those few years survivors would need to take precautions while outdoors to guard against the effects of ultraviolet rays, but the most serious effects would likely be a modest increase in skin cancer rates—perhaps on the order of 10 percent.

◇ How about the effects of fallout?

✦ Directly targeted countries and countries directly downwind would experience very high levels of fallout that would likely prove lethal to many people and animals—especially if no efforts were made to seek shelter before the wind-borne fallout arrived. Radioactive fallout is most intense during the immediate hours and days following a detonation, and the dose you would receive could depend greatly on what protective measures you took.

◇ But very few people have access to fallout shelters, and what about after they emerged?

✦ It turns out that even staying in a basement below ground can reduce your radiation exposure to one-tenth what it would have been if you were outside. If you came out after a week or two when radiation levels were much less, you would still risk the long-term effects of some increase in cancer incidence, but you would probably face many worse threats to your well-being than radiation. As they say, after a nuclear war, the survivors might envy the dead.

◇ How about the effects of nuclear winter?

✦ The first study to coin the term *nuclear winter* predicted that the world would be plunged to temperatures below freezing for many months, even if the war occurred in summer. More recent calculations show that the effects on the environment would probably be significant but not quite as serious as first thought, though large uncertainties remain. Some scientists have suggested the results might be more in the category of nuclear fall than nuclear winter.

◇ Maybe it would be more like a nuclear spring time. Let's hope we never perform the "experiment" to resolve the uncertainties!

Chapter 8

Earth

What if there were only one time zone?

✧   How many time zones are there now?

✦   There are 24—one for each hour Earth takes to rotate on its axis. That way, as the Sun passes overhead, it is roughly noon in the middle of a given time zone in most cases.

✧   What's the point of having time zones?

✦   The system got started in the 1870s following the widespread development of the railroads and the telegraph. Prior to then, each locality had its own local time. After all, if it took you a couple of days to travel a few hundred miles on horseback, who would care or even notice that different regions had their own local time?

✧   How did the railroads cope with the lack of standard time?

✦   Each railroad defined its own time relevant to a wide region. A railroad-specific time may have suited a railroad's purpose, but it meant that cities served by different railroads had different times for each one, as well as their own local time. In the Pittsburgh railway station you would have seen clocks reading six different times!

✧   Sounds like a mess. How was it resolved?

✦   The system developed by the railroads formed the basis of the standard time zones eventually defined for the United States. Shortly thereafter, an international agreement was reached to define time zones around the world.

✧   Why do the dividing lines between the time zones around the world have all those peculiar zigs and zags?

✦   That is based on economic and political factors according to the wishes of each state and country, which can decide for itself where the dividing lines should be drawn.

✧   Presumably, it would be advantageous for a country to be all in one time zone if possible.

✦ That's easy for a country like Chile, which, though it covers a long distance in latitude, occupies a small range in longitude.

✧ OK. What about a major large country like China?

✦ Funny you should mention China. The Chinese have opted to be all in one time zone, even though the width of the country spans a range of longitude that might warrant six different time zones.

✧ Doesn't that cause a lot of Confucian confusion? The time twelve noon might correspond to midmorning in one end of the country and midafternoon in the other end, based on the position of the Sun.

✦ You'd get used to that. But having a common time for the whole country has great advantages for internal communications and business. In fact, there might be considerable advantages for the world to adopt one time zone. For one thing, we'd get rid of the silly international date line, since the date would be the same for the whole globe at all times. You would still have to remember that people living on the opposite coast of the United States arose three hours earlier (or later), but at least the time would be the same everywhere.

✧ But wouldn't that mean that in some places people working nine-to-five jobs would have to go to work in the middle of the night?

✦ No, everyone could still work the same daytime or nighttime hours they do now; they just wouldn't call them "nine to five" in most places. There would be only one longitude where the Sun was overhead at the time called noon. On the opposite side of the planet, noon would occur in the middle of the night. In fact, that's probably why my sensible suggestion is not likely to be adopted. Everyone would want the one global time zone to agree as closely as possible with their current local time.

What if Earth didn't have a moon?

✧ It sure would get dark at night out in the country.

✦ Right. Probably, given the much darker nights, animals would have evolved to have somewhat better night vision—at least the predators.

✧ I think I'd have been one of those who made myself scarce at night. But other than werewolves and lovers, I can't think of too many others that would miss the Moon.

✦ Actually, the Moon has played a critical role in the development of life on Earth, because it provides a steadying influence that prevents Earth's axis from tipping chaotically. Without that influence, our seasons would vary in an unpredictable way. In addition, there were three times during the history of life on Earth when the presence of our Moon made a crucial difference to our evolution. The first time was when the earliest ancestors of land-based life first crawled onto land from the sea.

✧ What did the Moon have to do with that?

✦ The Moon is the main cause of the tides. With tides caused only by the Sun, the transition zone between sea and land would be much narrower. A narrower tidal zone would make it more difficult for creatures beginning to adapt to the land to spend part of their time in the sea.

✧ Sort of like having a set of training wheels, you could say. What was the second time the Moon's presence was crucial?

✦ When Isaac Newton published his theory of gravity in 1687.

✧ I thought it was Newton's being hit in the head with a fig—I mean an apple—that gave him the idea.

✦ Actually, it's said that Newton saw a falling apple with the Moon in the background, and he wondered, if the apple falls, why doesn't the Moon?

✧ OK, I give up. Why doesn't the Moon fall down?

✦ You could say that in a sense the Moon does fall, but its horizontal (orbital) motion is fast enough so that it never hits Earth.

✧ But why doesn't it hit Earth?

✦ Imagine an object thrown horizontally off a high tower on Earth. The distance from the tower base to the spot where it lands increases as the speed of the throw increases. If it were given enough initial speed, the object would never land (it would follow Earth's curvature), and, were it not for Earth's atmosphere, the object would be in orbit just like the Moon.

✧ Or just like an artificial satellite! But how come the Moon takes a whole month to orbit Earth, but artificial satellites orbit in under 90 minutes?

✦ Earth's gravity is a lot weaker at the distance of the Moon than on Earth's surface. In fact, Newton realized that the lunar orbit period of a month (actually 28 days) is exactly what it should be if the strength of gravity decreased in the way he thought it should.

✧ OK, what way is that?

✦ The inverse square law: since the Moon is 60 Earth radii away, the strength of Earth's gravity there is only $1/(60 \times 60)$ of its strength at Earth's surface, which can be shown to explain exactly the Moon's 28-day orbit.

✧ You said before that there were three times when the presence of the Moon made a crucial difference to life on Earth. What was the third?

✦ Why, when man first set foot on the Moon, of course! Without a moon to aim for, it is unclear if the space program would ever have gotten off the ground, so to speak. The nearest potentially habitable celestial object besides the Moon is Mars—roughly 200 times farther away at closest approach, and a much more difficult journey.

✧ So you could say that the Moon once again provided Earth creatures with "training wheels."

✦ Right! And if humans should ever venture far into space, it will be essential to have lunar bases, since "jumping off" into space from the Moon is a lot easier than getting off Earth, given the Moon's much smaller escape velocity.

✧ Did you ever think about getting a public relations job at NASA?

What if a day were a year long?

✧ How could that be possible?

✦ Easy. If Earth always kept the same side to the Sun, then it would complete one rotation (a day) in the same time it completed one revolution (a year). Imagine yourself warily circling around a chained growling dog, always facing it while you circled it, and you'll get the idea.

✧ So in that case Earth would be in a sense just like the Moon, which always keeps the same face toward us. Incidentally, how come the Moon does that?

✦ Just as the Moon's tidal forces slow down Earth's spin, Earth has done the same to the Moon. In fact, the Moon's spin has slowed to the point where it always keeps its same face toward us.

✧ So what are tidal forces, and how do they slow down a body's spin?

✦ Each part of Earth experiences a slightly different pull due to the Moon's gravity. The net result of all these pulls is to distort Earth's shape into an ellipsoid—which you can think of as a sphere with bulges on opposite sides. Clearly, the water on the planet bulges to a greater extent than the solid ground, and these tidal bulges are held in place by the Moon's gravity as the solid Earth rotates under them. Thus, as Earth rotates underneath the tidal bulges, someone at a fixed point on the planet would see a tidal bulge (high tide) move past roughly every half rotation, or twelve hours. The resulting friction between the rotating Earth and the tidal bulges slows the planet's rate of spin.

✧ So could Earth actually lose its spin to the point where it always keeps the same face toward the Sun or the Moon?

✦ Yes, but at the rate Earth's spin is now slowing down, you won't be around to see it. In fact, it probably won't occur during the 5 billion years the Sun has left in its hydrogen-burning phase. Also, Earth would eventually keep its same face toward the Moon, not the Sun, since the Moon has a greater tidal influence on Earth than the Sun.

✧ Now I know you don't know what you're talking about. Everyone knows the Sun's gravity is stronger than the Moon's, since we go around the Sun, not the Moon. How could the Moon be the main cause of the tides?

✦ Because it is not the absolute strength of gravity due to each body that counts, but the *difference* in that force on opposite sides of Earth. The difference in the strength of lunar gravity on opposite sides of Earth is roughly twice as great as the difference in solar gravity.

✧ OK, so maybe Earth will eventually keep the same face toward the Moon, but is there any planet that always faces the Sun?

✦ No. At one time it was thought that Mercury did. In fact, Mercury completes an orbit in a time exactly 50 percent longer than one rotation on its axis. But let's imagine a fictitious planet that always keeps the same face toward the sun.

✧ Let's also pretend Earth, like our fictitious planet, always kept the same face toward the Sun. What would conditions on Earth be like?

✦ The side facing the Sun would become blazingly hot, and the dark side extremely cold. The extreme temperature differences would drive continually circulating winds. Any liquid water initially present on the bright side would evaporate, and most of it would find its way to the dark side, where it would fall as snow and remain on the ground.

✧ With no water on the bright side and only ice and snow on the frozen dark side, things don't sound too promising for the existence of life.

✦ Actually, there would be a very narrow habitable zone between the bright and dark sides where temperatures varied from freezing cold at one edge to blazing hot at the other, and where liquid water would exist.

✧ That sounds a little like saying that my front facing a campfire is blazing hot, my back is freezing cold, but "on average" I am comfortable. Incidentally, would there be a water cycle? It seems as if the water would either flow toward the dark side and stay there frozen, or flow toward the bright side and evaporate.

✦ Actually, there would be a water cycle in the narrow zone between bright and dark sides. At one edge of the zone, melting glaciers would feed a river that flowed toward the bright side, where it would then evaporate, allowing the water vapor to be carried back to the dark side.

✧ That sounds like a really strange river to me—its length would be maybe a few hundred miles across the transition zone between the dark and light sides, and its "width" would be the circumference of Earth.

What if grapefruit-size hail were common?

✧ Presumably, people would need to wear helmets anytime they were outside. But I don't see how plants and animals could possibly survive under such conditions.

✦ You might be right about that. If we assume that grapefruit-size hail was common throughout the history of the planet, it would seem likely that plants and animals would have evolved to deal with the hazard. Plants, for example, which need to spread leaves out to collect sunlight, might evolve to retract or rollup the leaves into a sturdy stalk at the first sign of hail.

✧ How might animals have evolved to cope with the danger?

✦ Probably by evolving a tough turtlelike shell that could withstand the impacts. But other possibilities might also be imagined. Some creatures might evolve long, bladelike, pointed hard heads that would deflect or split the descending hail.

✧ I suppose we could even imagine several races of humans evolving together—the pointy-heads (intellectuals?) and the flatheads (plane folks?).

✦ Possible, but unlikely. One of these two solutions to the hail problem is probably superior, and if both types did exist, the superior one would probably win out in the process of natural selection (survival of the fittest), after a number of generations—depending on how big an advantage it had.

✧ Are there other ways that natural selection might shape a creature's anatomy in response to the hail danger?

✦ Very likely animals' heads would not be on top of their bodies—the place at greatest risk from falling hail. A likely place for the brain might be near a creature's midsection, or perhaps its sex organs, which are also in need of protection.

✧ I've known some people whose brains were . . . well never mind. But is the prospect of grapefruit-size hail a realistic possibility?

✦ Hail this size has occurred. Apparently, the required atmospheric conditions are rare. Presumably the hail, formed at high altitude, needs to fall a long way through air containing many small droplets that are just on the verge of freezing. The freezing water then adheres to the falling hail particle, continually enlarging it.

✧ But if grapefruit-size hail can form, how come we never see grapefruit-size raindrops? Why are all raindrops about the same size?

✦ Actually, raindrops come in a variety of sizes ranging from a fine mist to the large droplets of a cats-and-dogs downpour. But there does seem to be a largest-possible water droplet in a rainstorm. Have you ever noticed when observing a leaky faucet how droplets always break off when they reach a certain maximum size?

✧ Yes, but what has that got to do with a maximum-size raindrop?

✦ Droplets and bubbles try to keep a spherical shape due to the phenomenon of surface tension. Essentially, the water surface acts like a stretched rubber membrane that holds the bubble or droplet together. Bubbles pop and droplets split when outside forces become sizable. For a falling water droplet, the force of the onrushing air can be enough to split a big droplet.

✧ But why are big droplets much more likely to split?

✦ The bigger the water droplet, the faster its maximum velocity of fall, and the greater the force of the onrushing air. So at some point the force of air resistance is enough to distort and then split the droplet, despite the surface tension tending to hold it together.

What if Earth were hollow?

✧ Of course, I assume this one is completely hypothetical—right?

✦ Sure. Even if we somehow succeeded in creating a large hollow cavity inside Earth, the inward force of gravity would quickly lead to its collapse.

✧ How come? You can have large chambers inside caves. Why would it be impossible to have a chamber that was a fair fraction of the size of Earth?

✦ The larger the chamber, the more likely it would be to collapse—in just the same way that the longer an unsupported section of bridge, the more likely it is to collapse. Also, a chamber at the center of Earth would have the weight of all the layers of Earth above it—leading to pressures of millions of atmospheres. No material on Earth could resist such a crushing force.

✧ Not even kryptonite? Even though we couldn't actually create a large hollow cavity at the center of Earth, let's pretend that Earth were in fact hollow. Could we tell if it was?

✦ The removal an appreciable fraction of Earth's interior mass would reduce the strength of gravity on its surface, so we could easily tell if you waved a magic wand that caused Earth to become hollow. But if Earth was always hollow, we wouldn't be able to compare its old and new surface gravity, so we couldn't learn of its hollowness so easily.

✧ So how do we know Earth really isn't hollow?

✦ In the same way you learn about the inside of a watermelon at the food store—by thumping it. In a scientific version of a planetary thump, geologists create seismic waves by detonating explosives on the surface. By recording the time these waves take to reach detectors at various points around the globe, we can deduce any boundaries between different layers inside Earth. A hollow central core would not escape notice in such measurements.

✧ Granted that it wouldn't be possible to hollow out Earth, could we imagine hollowing out a large asteroid?

✦ Aside from big practical problems, such as getting there, hollowing out an asteroid would be a lot easier. Given the much smaller gravity on an asteroid compared with Earth, the inward force might not be enough to cause a large cavity to collapse.

✧ What would it be like inside a hollow asteroid?

✦   Imagine that the asteroid had the shape of a hollow sphere, and that you were at the very center. At that point gravity would be exactly zero. Otherwise, you could point in a direction that was "down," which would be impossible at the very center, where all directions would be equivalent. Surprisingly, it is not only at the very center of the hollow asteroid that the strength of gravity would be zero. You would actually be weightless everywhere inside the hollow asteroid!

✧   How could that be? Suppose you were located near one side—wouldn't that side attract you more than the more distant mass on the other side because of the inverse-square law?

✦   It can be shown by clever mathematics that the nearness of the mass on the side you are close to exactly offsets the greater amount of mass on the far side of the asteroid, with the net result that the gravitational pull in all

Failure to shut off the flow from the world's largest underground helium well leads to catastrophe.

directions is exactly equal, and hence cancels out. Imagine you were standing on the outer surface of the asteroid subject to its surface gravity. You could then open a hatch, climb inside, and literally float across to the other side in complete weightlessness.

✧ It certainly sounds like fun to float around weightless, but aside from recreational purposes, why would anyone even consider hollowing out an asteroid?

✦ A hollowed-out asteroid could be virtually a miniature world ideally suited for an orbiting space colony or for an extended space journey spanning many generations.

What if Earth's atmosphere were much thinner?

✧ How thin is the atmosphere?

✦ The atmosphere gets thinner and thinner with increasing altitude, becoming half as dense at roughly 7 miles up. For a really thin atmosphere, consider the planet Mars. At ground level its atmosphere has only one percent of Earth's atmospheric pressure.

✧ Could living organisms cope with a much thinner atmosphere than now exists on Earth?

✦ At about 15,000 feet altitude, where the pressure is about 60 percent its sea-level value, pilots in unpressurized aircraft have to use oxygen masks. Even at much lower altitudes, the ability to engage in sustained physical activity can be impaired.

✧ Presumably, if life evolved in a very thin atmosphere, it would (within limits) be able to adapt to whatever the pressure happened to be.

✦ Yes, and life also would probably modify the atmosphere as it developed, as seems to have been the case on Earth. Oxygen, the product of plant respiration, was probably not present in Earth's original atmosphere.

✧ What would the weather be like if Earth's atmosphere were much thinner?

✦ The weather would resemble that at very high altitudes now. Temperatures would be much cooler with less of an insulating atmospheric "blanket," and with extreme temperature differences between night and day.

✧ How about other environmental conditions?

✦ With a much thinner atmosphere, Earth's surface would be a more hazardous place in many respects. We would, for example, be exposed to higher levels of harmful ultraviolet radiation from the Sun, and cosmic rays from space. Meteorites would be less likely to burn up in the atmosphere and would impact

Earth with greater frequency. Even hailstones would impact the ground with much greater speed in the thinner atmosphere.

✧   Would there be other effects on everyday life?

✦   You would probably need a pressure cooker to do your cooking, because at a much-reduced atmospheric pressure water boils at a much lower temperature—too low to cook your food. Playing golf would be great fun, since you could hit your drives much farther and straighter—so you probably wouldn't have to worry about hooks and slices.

✧   How about effects on transportation vehicles?

✦   Cars would get much better gas mileage with the reduced air resistance, but birds and planes would have a difficult time staying aloft in a much thinner atmosphere, and you could probably forget about parachutes.

✧   Which would be no loss if there were no planes. What other effects would a very thin atmosphere have?

✦   Except for extremely flammable fuels, most other things that now easily burn might no longer do so, given the reduced oxygen levels.

✧   I guess it's fortunate that Earth's atmosphere is not much thinner. How come some planets, such as Mars, do have a much thinner atmosphere than Earth, and the Moon has none?

✦   The main reason is the strength of gravity on a planet or moon. You may have thought that everything that goes up must come down, but that's not so. Objects heading upward faster than a certain speed called the escape velocity will leave the planet for good. On planets or moons with low gravity, many of the molecules in the atmosphere escape into space because their speeds exceed the escape velocity.

What if Earth were always cloud-covered?

✧   I've lived in some places like that, but I guess you mean the entire planet perpetually covered with clouds. Aside from it being rather depressing, what difference would a perpetual cloud cover make?

✦   Cutting down on the level of sunlight reaching the surface would drastically affect Earth's climate, and hence its vegetation, though the direction of the effect would depend on the altitude of the cloud cover.

✧   What do you mean?

✦   Low-altitude clouds mainly reflect incoming sunlight, and keep Earth cooler. Very thin high-altitude clouds cut down on sunlight too, but they also act as a "blanket" that traps heat radiation from the ground, and they keep the surface warmer.

✧ Let's assume that the modification of Earth's climate was not so drastic as to prevent the evolution of life as we know it. In this case, what would be some other effects of a perpetual cloud cover?

✦ Probably, we would have a very incomplete picture of our universe. Not only might we not have any awareness of the stars, but we might not even know about the existence of the Sun and Moon—assuming the cloud cover was fairly thick. In fact, it seems likely that we would not even realize that Earth was round.

✧ That's ridiculous. Astronauts routinely travel into space and photograph the planet.

✦ Yes, but remember if Earth were perpetually cloud-covered and we knew nothing of other heavenly bodies, we would have little incentive to travel into space, and we probably would not have developed the basic physical theory making space travel possible. After all, Newton's law of gravitation was developed based on earlier observations of heavenly bodies.

✧ How could we be unaware of Earth's roundness? People can easily travel round the planet and directly see that it's round.

✦ Yes, but the earliest voyagers who sought to circumnavigate Earth had good reason to believe it was round, and they could even estimate its diameter. Without a well-founded belief in a round Earth, it seems doubtful anyone would have started on a voyage of completely unknown duration.

✧ How did we get our earliest estimates of the diameter of Earth anyway?

✦ Around 200 B.C., the Greek astronomer Eratosthenes used a very clever method. To get the idea, think of yourself standing on a really small planet, say 50 feet in diameter. At the equator of the planet the Sun would be directly overhead at noon, and you would have no shadow. As you walked away from the equator, the length of your shadow would grow, because the Sun would no longer be directly overhead.

✧ But how could the length of my shadow tell me the planet's diameter?

✦ Assuming the Sun was so far away that its rays were parallel, the distance you would need to walk on this planet to find a given change in the length of your shadow would be directly proportional to the planet's diameter.

✧ Couldn't we have figured out that Earth was round some other way?

✦ One well-known observation is that when tall-masted sailing ships return to port, their masts are seen before their hulls because of Earth's curvature. But without being able to rely on the stars for navigation, it is likely that sailing ships on a cloud-covered planet would never venture far from land. People living on a perpetually cloud-covered planet would probably remain in the dark in more ways than one.

What if they dug a tunnel through Earth?

✧   This one is strictly science fiction—right?

✦   Probably. At least if you wanted the tunnel to go right through the center of the Earth. Our deepest mine shafts only go down a couple of miles—a mere .05 percent of the way to the center.

✧   Wouldn't the tunnel have to pass through molten iron at Earth's core?

✦   The central core is actually solid iron, because of the tremendous pressure of millions of atmospheres, but the outer core is liquid, so it is difficult to imagine such a tunnel ever being built, given the heat and tremendous pressure at the core.

✧   Well, let's just imagine such a tunnel existed and that people could somehow be protected from the heat and pressure. What could we do with it?

✦   Visits to the other side of the world would then take just about an hour, and they would require zero expenditure of fuel. If you stepped into the tunnel you would fall straight down, accelerating all the way to the center of the Earth.

✧   What happens at the center?

✦   As you approached the center your acceleration would slow, and you would reach your maximum speed just as you passed the center. Right at the center "down" and "up" would switch places as far as you were concerned, and your speed would slow down through the second half of your journey. All provided free, courtesy of gravity. Of course, it would be necessary to wear a space suit or be in a sealed vehicle.

✧   It sounds like it has great potential for an amusement park ride. Why would space suits be necessary?

✦   The air in the tunnel would need to be pumped out so as to allow you to fall unimpeded by air resistance. If any air remained, on falling into the tunnel, you would oscillate back and forth about the center of the Earth with decreasing amplitude, until you eventually came to rest at the center.

✧   If this idea of a tunnel through Earth is strictly science fiction, are there any similar ideas that could be feasible?

✦   One would be a tunnel, not through the center of the Earth, but along a line connecting two points on Earth's surface. If we ignore the effects of friction, gravity would speed a train up for the first half of the trip and slow it down for the second half, just as in the case of a tunnel through the middle of Earth.

✧   How long would the trip take?

✦   That's the neat part. The trip would take just about an hour, no matter where on Earth the two end points were—as long as the tunnel was straight.

Earth   113

For the shorter tunnels, the smaller component of gravity along the tunnel direction would just compensate for the trip's reduced length, so the trip would always take the same amount of time. Unfortunately, we are talking about a hypothetical situation, because to achieve a constant-time trip for all tunnels, we would need to eliminate the considerable effects of friction and unrealistically ignore variations in Earth's density.

✧   I assume that just like the tunnel through the center of the Earth, these tunnels would have to be air-free.

✦   Actually, someone has suggested pumping out some of the air only on the front side of the train. That way, if the train makes a close seal with the tunnel, the greater pressure behind the train could propel it forward—possibly with enough force to overcome the losses due to friction. As with the tunnel through the center of the Earth, we'd have a free ride—courtesy of gravity, with an air pressure assist.

What if Earth's axis were tilted on its side?

✧   I'm not sure what it means to have Earth's axis on its side. Wouldn't that put the north pole at the equator, or something ridiculous like that?

✦   No. It only means that as Earth orbited the Sun, its axis, lying in the plane of its orbit and pointing in a fixed direction, would point directly at the Sun two times of the year.

✧   How would the Sun move across the sky each day if Earth's axis lay on its side?

✦   To help you visualize this, you might want to put a plate at the center of your kitchen table (representing the Sun) and place an orange on the table (representing Earth). As Earth orbits the Sun, its axis (passing through the navel) points in a fixed direction. If, at some point in the orbit, the axis pointed directly toward the Sun, the Sun would remain fixed during the entire day at a point directly over the north pole (the navel). At this time of year (midsummer), it would be blazingly hot at the north pole.

✧   What would the weather be like at other places on Earth at this same point in Earth's orbit?

✦   If Earth's axis were to point toward the Sun, the Sun would be in the location of the North Star now. The Sun would stay above (below) the horizon in the Northern (Southern) Hemisphere during Earth's entire 24-hour rotation. Thus, on such a world, in the Southern Hemisphere it would be extremely cold, and in the Northern Hemisphere it would be extremely hot. Right at the equator, the Sun would appear to go in a circle right on the horizon during the course of the day.

✧ How would these conditions change during the course of the year, as Earth revolved about the Sun?

✦ As is the case with the actual Earth, six months later the weather in the Northern and Southern Hemispheres would reverse, when Earth found itself on the opposite side of its orbit. The main difference from the actual Earth would be that the temperature extremes between summer and winter would be much greater.

✧ What causes these seasonal temperature changes?

✦ Think about catching falling raindrops in a tray. If the rain descended vertically downward, you'd catch much more rain in the tray than if it were blown by the wind and descended obliquely. The same is true when it comes to catching the Sun's rays, which descend vertically only when the Sun is directly overhead.

✧ So if Earth's axis were on its side, why would summers be so much hotter and winters so much colder than they are now?

✦ The reason summers are so hot now near the equator is because the Sun spends a greater fraction of the time high in the sky, so that its rays shine nearly directly downward—allowing more rays to fall on each square meter than at locations away from the equator. If Earth's axis lay in the plane of its orbit, at the north pole during summers the Sun would spend all its time above the horizon high in the sky, and it would be blazingly hot at the north pole.

✧ How would it have affected the evolution of life on Earth if summers had been much hotter and winters much colder?

✦ Probably the seasonal adaptations of plants and animals (such as shedding leaves and thickening fur) would have been more pronounced. Possibly, many more animals would have evolved to hibernate as a means of winter survival.

✧ How about the opposite extreme of the one we are considering here, if Earth's axis were exactly perpendicular to the plane of its orbit?

✦ In that case, there would be nearly no seasonal variation at any point on Earth. Picture yourself standing at any point on this perpendicular-axis Earth as it rotated. The Sun would appear to move across the sky along exactly the same path regardless of where Earth was in its orbit. The average angle at which the Sun's rays hit the ground would be the same at all points in Earth's orbit—the effect being almost no seasons.

✧ Why "almost"?

✦ If Earth's distance from the Sun varied a bit during the year, as is now the case, that would cause a small seasonal change, but not much. In the case of the real Earth, in the Northern Hemisphere, we are actually closer to the Sun during the winter.

✧ Are there any planets whose axis is actually on its side, and if so, how did they get that way?

✦ Yes, Uranus. Presumably, either the planet was "born" that way when the solar system was formed, or, more likely, its axis was flipped on its side as a result of an encounter with a large celestial body.

Chapter 9

Gravity

What if you lived on a small asteroid?

✧ You mean as in Saint-Exupery's wonderful story *The Little Prince*? That sounds like it would be a lot of fun. I'd love to hop around a low-gravity asteroid.

✦ Let's say you lived on an asteroid whose gravity was only a thousandth that of Earth, which would be the case for an asteroid having a diameter about a thousandth of Earth's—about 8 miles—except that an asteroid that size probably wouldn't be round.

✧ How come?

✦ You need to have a much larger heavenly body before gravity is strong enough to create enough inward force to crush the rocks and make it round. NASA flybys of comets and small moons show they are decidedly not round.

✧ What would life be like on the surface of a small asteroid?

✦ Since the asteroid could have a very irregular shape, the landscape would be very rugged. You would probably see few places where the horizon appeared flat. But due to the very low gravity, climbing the rugged landscape would pose little problem for you, since you would weigh only a thousandth as much, and you'd have to fall a considerable distance to get hurt.

✧ Wouldn't the very low gravity mean that you would be moving in slow motion?

✦ That's only in science fiction movies. You could move at your normal speed, except that falling objects would fall much more slowly. For example, if you dropped an object from chest height, it would take approximately 15 seconds to reach the ground.

✧ What would some other effects of the low gravity be?

✦ You could lift giant boulders (a thousand times more massive than on Earth), and you could jump a thousand times higher, assuming you could take off with the same speed as on Earth.

✧ What would walking be like?

✦ Walking would be more like flying, because each step you took would propel you forward a thousand times farther and higher than on Earth. But it could be tricky, because the slightest push against the ground would send you flying. If you took a step at a normal walking speed of, say, 2 miles per hour, you might slowly glide in an arc, taking you to a maximum height of perhaps 30 feet off the ground, and land 200 feet away, after spending a leisurely 90 seconds off the ground.

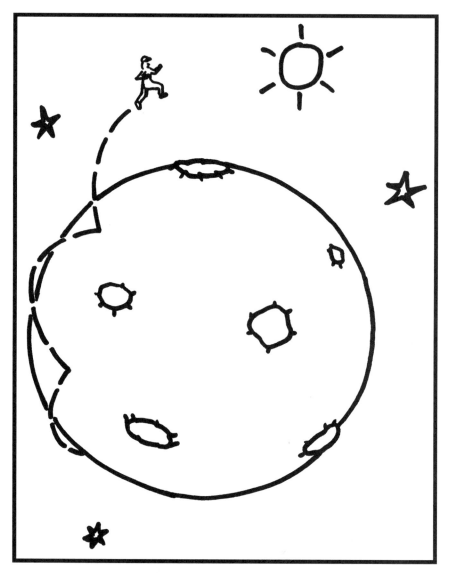

One small step for a man—one giant leap for mankind.

✧ Would you be in danger of jumping off into space and not falling back on an asteroid with a thousandth Earth's gravity?

✦ Probably not. On Earth, the escape velocity is about 25,000 miles per hour—which is the speed objects need to be hurled upward into space to escape Earth's gravity and never come back down. On the asteroid, the escape velocity would be a thousandth as great (25 miles per hour), which is faster than even a track star could run, let alone jump straight up.

✧ What would happen if I started running as fast as I could around the asteroid? Could I go into orbit?

✦ The speed needed to go into orbit is about 70 percent of the escape velocity. For the asteroid it would be around 18 miles per hour. On Earth, a track star could manage that for a short time, but on the asteroid you might need a bit of an assist to manage it—maybe a good pogo stick would be useful. Each hop would carry you farther and farther, until at orbital speed the surface would curve out from under you, assuming mountains didn't get in the way. The asteroid would certainly lack an atmosphere, due to its low gravity, so orbits at ground level would not be impeded by air resistance. To actually achieve orbit, you would need some additional thrust at the top of your jumps—a good fire extinguisher could do the trick.

✧ Great! Book me on the next flight.

What if the day were 85 minutes long?

✧ Why 85 minutes?

✦ Well, then Earth would be rotating 17 times faster than it does now.

✧ So what?

✦ At that rotation speed, the centrifugal force at the equator would be just enough for you to be weightless there.

✧ What do you mean "weightless"? Gravity would be just as strong as ever despite the faster rotation.

✦ Yes, but if you were standing on a scale your feet would not be pressing against the scale at all, because of the centrifugal force. You would literally be floating just above the scale.

✧ I've never been to the equator, but I find that difficult to believe. Shouldn't I feel some weight reduction at the equator now if Earth has to rotate only 17 times faster to make me weightless?

✦ Actually, you would feel a weight reduction now, but it would be only one part in 289 (17 squared), since the size of the centrifugal force depends on the square of the rotation rate.

✧ I see. So if I weighed 289 pounds, I would weigh one pound less at the equator, but if Earth spun 17 times faster, I'd lose the full 289 pounds.

✦ Right, but only at the equator. At the poles you would weigh the full 289 pounds, and at points in between, your weight reduction would depend on your latitude.

✧ I don't suppose it's an accident that a satellite in low Earth orbit completes 17 revolutions about the equator each day.

✦ No accident at all. With Earth rotating 17 times faster than now and you weightless at the equator you could gently push off, and you'd be in orbit just a foot off the ground! Well, better make that both feet off the ground.

✧ What would happen if Earth rotated still faster?

✦ In that case, the oceans and the atmosphere would be flung into space, and the very shape of the remaining solid Earth would be greatly distorted. Due to its high rate of spin, Earth would probably become extremely squashed, with its diameter at the equator appreciably more than that from pole to pole. At the equator you'd have to be tied down in order not to be thrown off.

✧ Why would objects fly off if Earth rotated fast enough?

✦ Some people would refer to "centrifugal force" to explain it. But we can just as well say that beyond a certain rotation rate the inward force of gravity is simply not great enough to keep objects traveling in a circle that would have Earth's radius.

✧ We're talking science fiction here—right? Earth couldn't really rotate that fast, could it?

✦ At one time, before tidal friction slowed it down, Earth did actually rotate much faster than it does now, but probably not 17 times faster. On the other hand, there is nothing to prevent a planet from rotating so fast that things would fly off. For example, in the case of a small asteroid where gravity is extremely weak, it wouldn't take much spin for you to get thrown off.

✧ I think I'd move to one of the poles if I lived on an asteroid; I don't like to be tied down.

What if gravity varied greatly on Earth?

✧ From the wording of the question, I guess we can assume that gravity does actually vary slightly with position.

✦ Right. As we just discussed, Earth's spin causes a small variation in your weight depending on your latitude. Also, the equatorial bulge causes a comparable variation. Finally, local underground concentrations of mass from mineral deposits also affect the strength of gravity at any given location.

✧ So how could we get really large variations in the strength of gravity at different points on Earth's surface? What about if Earth's iron core were located off center, perhaps?

✦ You would get a nonuniform strength of gravity if such a thing occurred, because places on Earth closest to the center of the iron ball would feel a stronger gravitational pull than points on the opposite side of Earth. But an off-center iron core would not be a realistic possibility. A massive spherical iron core located off center would automatically move to the center due to the gravitational pull of the rest of Earth.

✧ OK, how about if the iron were initially distributed throughout Earth's interior in a nonspherical way—maybe even in the shape of a giant bar magnet located along Earth's north-south magnetic axis?

✦ Earth's magnetic field does resemble the kind of pattern you would get if a giant bar magnet were stuck inside the planet, but no one really believes that the iron is actually distributed in the shape of a bar magnet. Such an arrangement would also be unstable, and the iron bar's own gravity would mold it into a spherical shape.

✧ So how could you get a large variation over Earth's surface?

✦ The only possibility would be if Earth were rotating much more rapidly than it does now. As we discussed, if Earth were rotating 17 times faster, you would be weightless at the equator and have your normal weight at the poles.

✧ What would life be like on an Earth in which gravity varied greatly with location?

✦ Plants and animals would undoubtedly evolve quite differently, depending on local gravity. At places where gravity was very weak, trees would probably grow much taller as they competed with one another for sunlight. Animals would probably have legs with much thinner bones. Likewise, people whose ancestors lived for a long time in one location would evolve in a way that best adapted them to the local gravity.

✧ I assume this means that people living near the low-gravity equator would tend to be taller with thinner leg bones than people elsewhere.

✦ Probably, but only if differences in the pull of gravity outweighed other influences on human evolution, such as climate—which itself would be drastically affected by varying gravity. At places where local gravity was weaker, the pressure of the atmosphere would be less, because gravity is what holds the atmosphere down. So animals would evolve a breathing apparatus that would be adapted to the local atmospheric pressure.

Gravity   121

✧   It sounds as if animals and people would look and function very differently at different points on Earth.

✦   Right. The differences could be so extreme that the "low-gravity skinnies" might not be able to visit those places on Earth where the "high-gravity chunkies" lived for very long. The chunkies might enjoy going to the lower-gravity equatorial environment for a vacation where they could eat to their heart's content, and simultaneously lose weight. But the much lower atmospheric pressure would be a real headache.

What if gravity gradually weakened with time?

✧   I guess that would be good news for weight watchers, who would lose pounds without dieting—even though they would look just as fat. Presumably, high jumpers also would be able to jump to greater heights, so it would be necessary to adjust high-jump records over time.

✦   You mentioned a couple of benefits, but there also would be a down side to a gradually weakening force of gravity. For one, it is the force of gravity that holds the Sun together against the outward pressure due to the heat and light generated at its core. With a reduced force of gravity, the Sun would expand, become less dense, and not burn as hotly.

✧   Presumably, that would lead to much lower temperatures on Earth. What other negative effects would a decreasing force of gravity lead to?

✦   Reducing the force of gravity that holds Earth in its orbit about the Sun would cause Earth to move into a larger-size orbit—which would further reduce temperatures on Earth.

✧   We are talking hypothetically here, aren't we? Surely there isn't any reason to believe that the force of gravity might actually be decreasing over time.

✦   This possibility was first suggested by the physicist P.A.M. Dirac. He also suggested a number of other crazy ideas, some of which have turned out to be true, and one of which got him a Nobel Prize. So we might want to take his "crazy" idea seriously.

✧   What made Dirac think gravity might be weakening over time?

✦   He observed that the strongest known force—the force that holds the particles of the atomic nucleus together—is known to be a "fortillion" times as strong as gravity. A "fortillion" is a name I just made up for the number written as a one followed by forty zeros, usually written as 10^{40}. Dirac also observed that when the age of the universe (about 10 billion years) is expressed in the units of time it takes light to cross the diameter of a proton, the result is also a "fortillion" such units.

✧   So I suppose that also means you would need to have a row of a "fortillion" protons in contact with one another to stretch across the universe. But why couldn't it just be a coincidence that these two numbers are equal?

✦   Dirac didn't believe in coincidences when it came to very large numbers. He reasoned that if gravity is now a fortillion times weaker than the strong nuclear force when the universe is a fortillion units old, maybe the relative strength of

the two forces was always proportional to the age of the universe at any given time.

✧ I assume that would mean the two forces would have been equally strong when the universe was one unit of time old and had a diameter equal to that of a proton. But couldn't we easily tell if gravity were actually weakening with time?

✦ If the strength of gravity varied inversely with the age of the universe (10 billion years), it would weaken by only one part in 10 billion each year.

✧ So I guess I had better not give up my dieting plans. But surely some clever scientist has figured out a way to test this idea, even though the effect is small.

✦ Yes, as a matter of fact. Thanks to some reflectors the astronauts left when they visited the Moon, we can bounce a laser pulse off the Moon and measure its distance with high precision. In round numbers, the distance to the Moon is about 10 billion inches. If gravity is weakening by one part in 10 billion each year, we could expected the average lunar distance from Earth to increase by around one inch per year.

✧ OK, so what do we find when we measure the average lunar distance?

✦ It is in fact increasing by more than one inch per year. But most of the increase is probably due to tidal forces between Earth and the Moon: as one body (Earth) slows its spin, the other body (the Moon) must revolve faster and go into a larger orbit. Unfortunately, the measurements are not accurate enough to tell whether the rate at which the Moon is receding is consistent with a constant force of gravity.

✧ Somehow, I knew this was going to be another one of those "we don't really know for sure" essays.

What if gravity shields existed?

✧ I always thought it would be nice to go for a ride on a magic carpet, which an antigravity shield would make possible.

✦ Yes, it would make for some interesting trips. The idea of a gravity shield has been used in any number of fantasy voyages, including Jules Verne's classic *From the Earth to the Moon*.

✧ How do we know there might not be some yet-to-be-discovered substance that could function as a gravity shield?

✦ Unfortunately, that does not seem possible. All matter seems to be affected by gravity in exactly the same way, as far as we can tell.

✧ What about that stuff called antimatter? Doesn't it have properties that are opposite to that of ordinary matter? Perhaps antimatter is repelled by the force of gravity rather than attracted.

✦ Even antimatter should fall down rather than up, but until an experiment is done using antimatter, we won't know for sure. It's not an easy experiment to do, because antimatter can be produced only as subatomic particles in high-energy accelerators. Also, it is difficult to separate the effects of gravity from the much larger electrical forces acting on tiny particles of antimatter.

✧ How come we can't shield against gravity, when we can shield against virtually anything else: heat, light, sound, radioactivity, magnetism, and electricity?

✦ The explanation is a little different for the first four and the last two items on your list. For the first four, it is possible to find materials that act as absorbers or reflectors of the particular form of energy, but there is no such thing as a gravity absorber or reflector.

✧ How about a black hole?

✦ Actually, a black hole might qualify as a gravity absorber, in the sense that by its presence it would alter the force of gravity between any two objects on opposite sides of it, but it certainly couldn't be used as a gravity shield in view of the tremendous gravity it itself exerts.

✧ How come we can shield against electricity and magnetism?

✦ In both cases, shielding depends on creating induced electric charges or currents that cancel out the effects. For example, if you cover the walls of your room completely with aluminum foil or another conducting material, you can keep any radio or other electromagnetic waves out, because induced currents on the surface of the walls will cancel out the radio signals inside your room. Try wrapping a portable radio in aluminum foil while it is playing and see what happens.

✧ How come this happens with electricity and magnetism, but not with gravity?

✦ To shield against electricity or magnetism, it is essential that electricity come in both positive and negative charges, so that both attraction and repulsion are possible. With gravity, there is only one kind of (positive) mass, so only attraction occurs, and no cancellation can occur.

✧ It sounds as if a universe in which the force of gravity were repulsive might make an interesting *what if* to talk about.

✦ Actually, you'd find it a pretty repulsive place, since no planets, stars, or galaxies could form, and individual atoms would fly apart forever.

What if gravity were not an inverse-square force?

✧ Maybe we better define our terms. I don't know what a square force is, let alone an inverse-square force.

✦ By an inverse-square force, we mean that the force between any two masses varies with the inverse square of their distance. Double the distance between two objects, and the force is 1/4 as much; triple it, and it's 1/9 as much, et cetera.

✧ In that case, why don't I notice a big decrease in the force of gravity holding me to Earth when I climb a stepladder?

✦ Because, it's the distance between you and the center of the Earth that counts, and that hardly changes when you climb the ladder.

✧ I've always had trouble thinking about gravity as somehow reaching across space, grabbing me, and pulling me down.

✦ You're not the only one. Think of the force of gravity between two objects as arising from some invisible spring connecting the two bodies—but with a very important difference. The force needed to stretch a spring is proportional to the amount of stretch, but in the case of the "gravity-spring," the force varies as the inverse square of the separation distance. A gravity-spring connecting two masses is more like a piece of taffy, where the force gets weaker as it stretches out.

✧ Aha—the taffy theory of gravity. Now I see the real meaning of the inverse-square law. So what would the consequences be if we lived in a nontaffy universe?

✦ Other than the inverse-square force, the possible kinds of forces that would permit planets to exist in closed stable orbits about the Sun are rather limited. In fact, the only two forces that allow elliptical orbits are the inverse-square force and a force that varies in direct proportion to the distance.

✧ I don't understand why stable orbits couldn't exist for, say, an inverse-cube law of gravity. Let's imagine a spring connecting two bodies as the lighter one orbits the heavier one in a circle. The force in the spring will be whatever it needs to be in order to keep the lighter body in its circular orbit. As long as the bodies stay a fixed distance apart, how could it possibly matter how the force in the spring varies with distance?

✦ Bravo. You are absolutely right. Circular orbits are possible for any type of force. But if you introduce the slightest disturbing influence (such as the presence of a third body anywhere in the vicinity), the orbit is no longer closed. For an inverse-cube force, for example, if the speed of the orbiting body is ever so slightly different from what it needs to be for a circular orbit, the body will spiral either inward or outward; the orbit is unstable.

✦ How about the inverse-square force? What happens to it if the orbiting body has a greater or lesser speed than what is required for a circular orbit?

✦ The body then travels in a stable elliptical orbit. As noted above, the only other force giving elliptical orbits is one where the force varies directly in proportion to the distance—the so-called spring force, because that's approximately how a spring behaves. Out of the infinite number of possible variations of gravitational force with distance, only these two types allow for stable elliptical orbits.

✦ Could the universe have had "spring gravity" instead of "taffy gravity"?

✦ Probably not, because stars and solar systems could not form in such a case. Stars form when a large region of hydrogen gas with slightly higher density than elsewhere collapses under its own gravity. To get a gravitational collapse, the strength of gravity must get stronger, not weaker, as the cloud contracts.

✦ So what would a "spring gravity" universe look like?

✦ Probably, it would be a collection of particles permanently bound to one another in continual oscillation, with no formation of stars, planets, or galaxies. Perhaps this was one of God's early experiments.

✦ Nothing to write a Book of Genesis about.

Chapter 10

Material
properties

What if all surfaces
were much more
slippery?

✧   In a superslick world, I assume it would be as if we were always standing or walking around on ice. Incidentally, why do you need to take such tiny steps when walking on ice?

✦   When you take a step, you push against the ground to move forward. The larger your step, the bigger the friction force on the bottom of your foot needs to be to propel you forward. If you take too large a step, there is not enough friction to keep your foot from sliding backward out from under you. Most people realize this instinctively after a few slips, and adjust their step size accordingly.

✧   Well, suppose the floor were even more slippery than ice? How slippery would it have to be for walking to be impossible?

✦   Once the friction force between the floor and your feet dropped below about half a percent of your weight, walking or even standing would be impossible, unless you were wearing shoes with cleats. Your shoes would also need to have special kinds of fasteners to stay on your feet.

✧   How so?

✦   Because knots rely on friction to stay tied, so your shoelaces would probably untie by themselves, and if you wore loafers, they might slide off your feet every time you took a step. For that matter, other kinds of clothing held up by friction, including socks, pants, and gloves might also slide off by themselves, but shirts and dresses would pose no problems.

✧   I guess pants could be tricky, depending on your anatomy. In what other ways would everyday life become more complicated?

✦   You might need special tools to pick things up. Just think of the trouble you have had picking up an extremely slippery object, and you'll see what I mean. Of course, the shape of the object would be a big factor: you could probably pick up an egg only by holding it in the palm of your hand. But getting an ultraslippery hard-boiled egg into the palm of your hand could prove rather tricky.

✧ How about other everyday items?

✦ Nails and screws also rely on friction, and they might not work either—even bolts held on with nuts. Most means of ground-based transportation would also have to be modified. Cars would need special tires to cope with icelike road surfaces, and brakes (which also rely on friction) would also need to be replaced—perhaps with rockets.

✧ Would there be any advantages to a world in which all surfaces were much more slippery?

✦ Yes, quite a few, though probably not enough to make up for the problems. If friction were negligible, cars and other vehicles might get much better gas mileage, because internal engine friction would be much less and also because they could glide along at constant speed with much less engine power—though some power would still be necessary to overcome air resistance.

✧ How about other means of transportation?

✦ If you had some way of sitting on a sled without sliding off it, you could coast for great distances on a level, nearly frictionless surface.

✧ Of course, in practice, the world has lots of hills and valleys, and you don't always want to go in straight lines.

✦ Of course, our entire discussion assumes that humans could exist if the world were a much more slippery place. In reality, it seems likely that evolution probably would have favored a very different course—perhaps favoring the development of flat, nearly two-dimensional creatures that slid around their world, propelling themselves by the expulsion of gas.

What if solid objects could pass through each other?

✧ There would be no more parking problems if a bunch of cars could all park in the same spot. But aside from ghosts walking through walls, are there any examples of objects capable of passing through each other?

✦ I can think of three cases that might qualify. The first would be that of ball lightning. This very rare weather phenomenon consists of a glowing ball of plasma that has on occasion been reported to pass through solid walls.

✧ Sounds pretty twilight zonish to me. What are the other cases?

✦ Here's one you can observe for yourself. Tie a couple of hammers at the ends of a piece of bare wire and drape the wire over an ice cube with the full weight of the hammers pulling the wire into the ice cube. Over a period of hours, as the weight of the hammers pulls the wire into the ice, the pressure causes the ice to melt beneath the wire. Since the water refreezes as the wire passes, the net result is an intact ice cube after the wire passes through it.

✧ That's a bit of a cheat, since you could say the wire leaves a hole in the ice that seals up as the wire goes through it. What's your other example?

✦ You probably won't like this one either. The other example is when two galaxies collide and pass through each other—an event astronomers can actually observe, but in extreme slow motion due to the long timescale involved. In a typical galaxy, the average distance between stars is roughly 30 million times their diameter. If stars were the size of grains of sand, their average separation would be around 20 miles. With such a large separation, it is not surprising that galaxies sometimes can pass through one another without stars colliding.[1]

✧ You're right, I don't like this one. Given the distances between stars compared to their size, it hardly seems reasonable to consider a galaxy to be a solid object. It seems more like a gas—the individual stars being like the molecules of a gas.

✦ You have a point. On the other hand, when you consider the amount of empty space in an ordinary solid object, you might expect that the atoms of one object should slip past the atoms of another in exactly the same way as the stars in a pair of galaxies. In a solid, the distance between atoms is roughly 100,000 times the diameter of the atomic nucleus. You might think that atoms should therefore just pass right through each other with so much of "solid" matter being empty space.

✧ OK, so what gives matter its solidity, if so much of it is empty space? Is it because electrons repel each other due to their electric charge?

✦ No, remember that atoms as a whole are uncharged. The solidity of matter arises because electrons and other subatomic particles obey something called the Pauli exclusion principle. At the atomic level, we don't think of matter as consisting of little hard spheres. Instead we describe its atoms in terms of a "wave function" that tells us where the electrons are likely to be found. The exclusion principle, when applied to two electrons, tells us that they don't like to get near each other. This effectively gives rise to a repulsion between neighboring atoms that prevents their interpenetration, despite all the "empty space."

✧ I'm not sure I understand this exclusion principle business, but suppose I

---

1. Possibly some stars might collide, given the large numbers of stars in each galaxy. Yet even if no stars collided, the galactic collision would not leave the galaxies unaffected. The distribution of the stars in each galaxy would be profoundly altered by gravity.

waved a magic wand over you, and the exclusion principle no longer applied to the matter of your body.

✦ In that case, all the electrons of my atoms would fall to their lowest energy state, close to the nucleus, and all the atoms would flock together, reducing my size to less than a dust particle.

✧ What would happen to you as a human dust particle?

✦ My weight would rest on two feet whose area was 10 billion times smaller, giving rise to a pressure on the ground 10 billion times larger than now. The ground would probably give way under such a large pressure, and I would fall completely through Earth, oscillating back and forth about the center with smaller and smaller amplitude—eventually coming to rest there.

✧ Suppose we extended the magic wand to include Earth as well?

✦ In that case, Earth itself would collapse inward under its own weight, and it would form a black hole the size of a golf ball with the entire mass of Earth.

What if water didn't boil?

✧ I assume that without a boiling point, water would simply get hotter and hotter as its temperature was raised. Why does water boil anyway?

✦ In water or other liquids, attractive forces between molecules keep them very close to one another, even as they collide and slide past each other. As the temperature is raised, the collisions at some point become violent enough to break the bonds between molecules. The water becomes a vapor, and bubbles are created.

✧ Presumably that is the point we call 100 degrees on the Celsius temperature scale.

✦ Right, but water boils at 100 degrees only when the outside pressure is one atmosphere. At altitudes where the pressure is lower, the boiling point is lower.

✧ Why should that be? The forces between water molecules are just as strong regardless of the outside pressure.

✦ If the outside pressure is higher, any small bubbles that begin to form are crushed, and the water needs to be raised to a higher temperature to allow the internal bubbles to grow and escape.

✧ What circumstances would be necessary for water to be incapable of boiling?

✦ For water not to boil, the attractive force between water molecules would have to decrease much more slowly with their separation than is actually the case.

If the force decreased slowly enough, no matter how violently a pair of molecules collided, there would not be enough energy to separate them.

✧ OK, let's imagine this were true. What would be some of the consequences?

✦ If the force between water molecules didn't allow it to boil, water wouldn't evaporate either. Evaporation occurs when molecules with speeds that are faster than average break free from a liquid's surface. The molecules left behind are slower than average, and the liquid cools as a result.

✧ I always wondered why blowing on a spoonful of soup caused it to cool. I imagine that by blowing away the newly escaped soup molecules, I clear the air and allow other molecules to evaporate, thereby speeding up the process.

✦ Exactly. But if we were to assume that water couldn't boil and therefore couldn't evaporate, there would be a lot more serious consequences than not being able to cool your soup by blowing on it. For example, all warm-blooded animals use evaporation of perspiration to regulate their temperature. Without evaporation we might all be cold-blooded reptiles.

✧ I think I'd be a turtle. What other effects would there be if water couldn't boil or evaporate?

✦ Without evaporation, the water cycle as we know it could not exist. The way it works now is that heat from the Sun evaporates water, which later comes down as precipitation. With no evaporation, there would be no precipitation.

✧ It might be nice if the sky were always free of clouds and rain.

✦ Actually it would be a disaster. Rain cleans the air, which would become much dirtier than now. Also, without rain to feed streams and rivers they would shortly dry up, leaving all the water to pool up in large lakes and oceans.

What if things didn't burn?

✧ Doesn't everything burn if the temperature is high enough?

✦ Not at all. Inert gases like helium don't burn, and certainly water and many other substances don't burn.

✧ OK, I've got a burning desire to know—what is burning anyway?

✦ When a substance reacts with oxygen so as to produce heat, we call that reaction "burning," or combustion. The fire that usually accompanies burning is due to the fact that the energy released in the reaction is sufficient to raise atoms to higher energy levels, thereby producing light and heat.

✧ Exactly how does combustion work? Let's say I burn some coal in the presence of oxygen. Where exactly does the energy come from?

✦ Coal is mostly carbon, which we'll represent as C. At room temperature, when a molecule of oxygen (O_2) smacks into a carbon atom, it just bounces away due to the slight repulsion between them. But if you heat the coal up beyond its kindling temperature, the two molecules collide more violently and approach each other more closely. At the closer distance they attract rather than repel, and their attraction causes them to rush into each other's embrace: C plus O_2 combine to form CO_2. Stored energy is released in the process, owing to the violence of the collision.

✧ It sounds very much like the way energy is released by a mousetrap when a Ping-Pong ball is dropped on it. Presumably, the coal-burning reaction becomes self-sustaining because the violent collision causes more energy to be emitted than the molecules had before they rushed to meet each other. But how come burning doesn't cause a chain-reaction explosion?

✦ Sometimes it does, particularly with gases or very volatile liquids. But when solids burn, the burning often takes place only on the surface, where oxygen molecules can react with those of the fuel. Even with solids, however, burning can produce an explosion if the fuel is oxygenated, allowing the reaction to spread very rapidly through the fuel.

✧ How did the energy stored in fuels get there in the first place?

✦ It ultimately got there as a result of energy given out by the Sun and stored by plants during photosynthesis. Some of these plants were eaten by animals, and the fossils of both were converted by heat and pressure into the fossil fuels we use today, millions of years later.

✧ So how would it be possible for things like coal and gasoline not to burn?

✦ I can think of only two plausible ways. One possibility, admittedly a bit of a cheat, is if everything had too high a kindling temperature to be ignited by ordinary natural means. Metallic aluminum, for example, burns but at such a high temperature that you wouldn't normally be aware of it.[2]

✧ What's the other way things might not burn?

✦ The other possibility is if the energy released during burning were sufficiently small so that you would never have a runaway chain reaction, and burning never produced any fire.

✧ Like the way my body "burns" off calories when I exercise. What would life be like if there were no such thing as fire?

✦ The discovery of how to start a fire was one of humanity's all-time biggies. Without fire, humans would probably never have learned to cook food or refine metals, and may never have had the warmth to be able to spread out beyond

---

2. Because of the high temperature at which metallic aluminum burns, producing aluminum metal from its ore was extremely difficult until modern times—so much so that aluminum was at one time more valuable than gold.

the equatorial areas of the globe to colder climates. It is highly doubtful whether humans could have bypassed the low but essential technology represented by fire and reached today's high technology of microwave ovens and electricity. Without fire, we probably would not have developed a civilization.

What if water's surface tension were much greater?

✧   My doctor told me not to be too concerned about the headaches I've been having, since they were just due to *surface* tension. OK, what's surface tension?

✦   Surface tension is what allows certain bugs to walk on water without breaking through the surface. In fact, you can easily demonstrate surface tension for yourself using razor blades or needles. Even though they are heavier than water and will sink if dropped into water, razor blades or needles placed gently on the surface in a horizontal position will actually float.

✧   I assume that only very light creatures like bugs can walk on water. Surely heavier creatures would sink.

✦   There is a species of small frog known as the "Jesus Christ frog" that manages the trick of hopping on water. But it would probably break through the surface when standing still. The frog gets extra support from the water by smacking into it, in much the same way that a thrown stone can skim the surface of the water but sinks when stationary.

✧   How does the water surface support heavier-than-water objects like razor blades?

✦   The surface of water and other liquids behaves as if it consists of a membrane that requires a certain force to tear it apart. If you think of a length of the water membrane as being stitched together, surface tension is the amount of force per unit length that the "stitching" would need to exert to prevent tearing.

✧   Why does surface tension exist, anyway?

✦   It comes about because all water molecules attract their nearest neighbors. Think of a bunch of people, all "magnetically" attracted to their neighbors. They would form a crowd packed as tightly as could be. If you poked the crowd with a pole, the pole would have to force apart adjacent people, and it would take some work to break through the "surface" of the crowd.

✧   How come we never hear anything about "interior tension"? Once your pole is entirely within the crowd of "magnetic" people, doesn't it still take work to pry people apart when the pole moves?

✦   For every pair of people the pole pries apart, there will be another pair at the other end of the pole that move together when the pole gets out of their way. In fact this other pair of people who come together supplies the force to push the pole into the first pair, so that the net force on the pole is zero when it is entirely within the crowd.

✧ What are some other consequences of surface tension, and what would the world be like if it were much greater?

✦ Surface tension is what causes water droplets and bubbles to be round, as all the water molecules (think "magnetic" people) try to get as close as possible to one another. The larger the surface tension, the bigger water droplets could get before other outside forces broke them apart. In a world where surface tension were much larger, raindrops would be huge.

✧ I imagine that if huge water droplets adhered to my body after taking a shower or as a result of perspiration, they would increase my weight appreciably.

✦ That's exactly right. You might be in a situation similar to that in which mice are now. A wet mouse actually doubles its weight when it becomes wet. For many insects the situation is, of course, even worse, since they can drown in a single drop of water.

✧ What would be some other effects of greatly increased surface tension?

✦ Well, you must have tried that trick where you stick a straw in a glass of soda, then cover the top with your thumb and lift the straw with the soda inside. It doesn't pour out until you release your thumb, because of surface tension at the bottom. If surface tension were much greater, you could do the soda straw trick with very much thicker "straws." In fact, you could probably turn a glass full of soda upside down and not have it pour out. Pouring liquids might become difficult or impossible in such a world.

✧ It sounds like only really big suckers could use straws in such a world.

What if air resistance were much larger?

✧ In that case, I suppose, air wouldn't get out of our way so easily.

✦ That's quite right. Depending on just how large air resistance was, walking through air might be just like walking through water, or conceivably even walking through molasses.

✧ I would assume that if air resistance were much larger, the effects on everyday life would be profound. How could people move about?

✦ Very slowly. The force of air resistance becomes more important the higher the velocity. If air were a much more resistive medium, it would take great effort to move through it, unless you did so at very slow speed.

✧ Could we swim through the air, as we now swim through water?

✦ That would depend on just how dense the air was. It is conceivable that if the air had a density comparable to the atmosphere on Venus (about a tenth that of water), a physically fit person could strap on a pair of wings and fly.

✧ Is density what determines how sizable air or water resistance is?

✦ Only in part—the other factor is viscosity. Water has a similar density to molasses or tar, yet it is much easier to move through. We normally think of gases, liquids, and solids as being distinct states of matter, but there are materials that are in between. Tar, for example, can be thought of as a liquid that is in the process of solidifying.

✧   If air resistance were much larger, how would that have affected the evolution of life on land?

✦   Undoubtedly, land creatures would have evolved to look much more like their sea cousins. Streamlining of body shapes would have been much more important in our evolution. Very likely, we would not have evolved to have such a wide front, and our method of locomotion would also probably be quite different—possibly even fishlike. Breathing would also be a major problem, because air could not easily be forced through narrow tubes. We might have evolved to get oxygen through gills rather than lungs.

✧   Any other implications of having a much greater value of air resistance?

✦   You might not have to be overly concerned about falling off cliffs. With a very high air resistance, a falling object would not attain a very high velocity—or terminal velocity—even if it fell a great distance. The terminal velocity attained by a falling object depends on its size and shape as well as the amount of air resistance. Given the air resistance in the actual atmosphere, the terminal velocity of a mouse is such that it can fall down a mine shaft and not get hurt, but a person would be killed, and a horse would splatter.

Chapter 11

Constants
of nature

What if the speed
of sound in air were
4 miles per hour?

✧   Is there any particular significance to the choice of 4 miles per hour?

✦   No—it's just the speed of a fast walk, so you'd be able to break the "sound barrier" while running. There is, of course, no problem in exceeding the speed of sound. That "barrier" business is just some terminology from the days before supersonic flight was common, when some misguided people thought there might be a barrier.

✧   So what would be some of the consequences of the speed of sound being only 4 miles per hour?

✦   Four miles per hour is equivalent to covering a distance of about 6 feet in one second. If we were having this conversation while standing 3 feet apart, there would be a one-second delay between each part of our conversation due to the round-trip travel time. If you have ever spoken on a phone line that introduced a fraction of a second time delay, you'll appreciate the strange effect that even a short delay can have.

✧   I suppose normal conversations would be very difficult to conduct if the speakers were much farther than 3 feet apart.

✦   On the contrary, in a world where the speed of sound was very low, it would become instinctive to allow for the travel time of sound. But communicating by voice would have some definite limitations. You could probably forget about shouted warnings like "duck!" which would probably arrive too late to do any good, and car horns would be useless.

✧   What would I hear if I were walking away from a source of sound?

✦   If you walked under 4 miles per hour, you'd hear the sound Doppler-shifted—just like the shift in pitch or frequency you now hear from a siren on a moving police car, only the effect would occur at much lower speeds. If you walked at 2 miles per hour, for example, all sound frequencies would be halved.

✧ Suppose I walked away right at the speed of sound?

✦ In that case all frequencies would be reduced by 100 percent, which is another way of saying that you wouldn't hear anything. You'd be staying even with the sound wave, and it would never reach you.

✧ Since there is no real sound barrier, what would happen if I walked or ran away from the source of sound faster than sound's speed?

✦ In that case you'd hear the sound in backward sequence, because you'd be continually overtaking sound waves that started out at earlier and earlier times.

✧ I wonder if I'd hear any satanic messages. How about if, instead of my moving faster than the speed of sound, it was the source of sound that was moving at supersonic speed?

◆   If the source were approaching you, you'd again hear the sound in backward time sequence. If it were not heading directly toward or away from you, you'd hear sonic booms just as you hear now when a supersonic jet aircraft flies overhead. A sonic boom occurs because, in this case, the sounds created at all different points along the object's path reach the listener at a single instant.

✧   Is there any conceivable way the speed of sound could be reduced to anything like 4 miles per hour?

◆   The speed of sound in air does decrease with decreasing temperature, but even if you reduced the temperature close to the point where air liquefies, the speed of sound would only be reduced to perhaps half its room-temperature value of 1,100 feet per second. There are other gases that have lower sound speeds than air, but 4 miles per hour is just science fiction.

What if the speed of light were 10 miles per hour?

✧   I guess in that case if I drove at 60 miles per hour I would be going faster than the speed of light.

◆   Not really, because we shall assume that the speed of light continues to be a "universal speed limit," as is now the case.

✧   I shouldn't say this, but I sometimes do drive faster than the speed limit. Why couldn't I exceed the speed of light if I kept accelerating my car at a constant rate of, say, one mile per hour each second? I would think that at that rate I could reach 10 miles per hour, the assumed speed of light, in 10 seconds of acceleration. Changing the speed of light isn't going to change my car's horsepower, is it?

◆   Your car would have as much horsepower as usual. What you would find is that your speed would initially increase in the expected way (one mile per hour each second). But as you began to approach 10 miles per hour, you would notice "diminishing returns": your speed would increase by a smaller and smaller value each second. As a result, your car's speed would never exceed 10 miles per hour, no matter how long you accelerated.

✧   I think I'd take my car back to the dealer if that happened. How could that possibly be?

◆   One way to explain it is to say the mass of your car would increase as you approached the speed of light, so that your car would become harder and harder to accelerate. In the real world we find it difficult to accept the idea of an upper limit to speed, because all our experience is with objects moving at speeds much less than the speed of light.

✧   I see. But if the speed of light were only 10 miles per hour, it would be part of our "commonsense" everyday experience that 10 miles per hour were the upper limit to speed. We might no more question the existence of a highest possible speed than we now question the existence of a lowest possible temperature.

✦   Right, and all the other strange effects predicted by Einstein's theory of relativity would also be part of our everyday "commonsense" world. The length of the car, for example, would be observed to contract along its direction of motion, according to someone watching it move past—its length decreasing to zero as its speed approached the speed of light.

✧   This sounds very peculiar—the car would simultaneously shrink and get heavier?

✦   Not heavier, which refers to its weight—just more massive, which means it would be harder to accelerate.

✧   What would the driver think of all the increase in the car's mass?

✦   The driver could consider the car to be at rest, and he wouldn't find anything abnormal about its mass and length. In fact, for the driver, the rest of the world outside the car would be moving past and looking all contracted. Passing pedestrians would look tall and skinny if standing or short and squat if lying down, since the contraction would be only along the direction of motion.

✧   But presumably, these effects are all just apparent changes; there is no "real" change in an object's length or mass . . . right?

✦   Actually, you might be hard put to define a "real" length or mass. We can measure an object's length when it is at rest in front of us, but such a measurement is no more "real" than a measurement made of an object's length as it moves past.

✧   Have any of these crazy effects actually been observed, or is relativity just a theory?

✦   Yes, they have all been observed. Even though the effects are small at ordinary speeds, accurate measuring devices have shown all these effects to occur: the speed of light as a universal speed limit, the increase in mass with speed, and even length contraction, as well as others.

✧   OK, what others?

✦   Well, a particularly interesting one is known as time dilation, which you could say means that moving "clocks" run slow—interpreting clocks in the broadest sense.

✧   So if I were in a high-speed spaceship, someone watching me go past would see everything happening in slow motion inside the ship.

✦ Right. But inside the ship, you wouldn't notice anything strange.

✧ I get it, because as the spaceship rider I could consider myself at rest—right?

✦ Yes, but someone watching your ship go past would have an equally good explanation: you wouldn't notice anything because your brain was running slow, too!

What if the speed of light were a million times greater?

✧ The speed of light is actually so high now, at 186,000 miles per second, I don't see how it would matter if it were 186,000 million miles per second.

✦ Actually, it would matter a great deal. To begin with, the speed of light could not have been measured until fairly recently, when sufficiently accurate instruments became available. More importantly, the technology allowing those measurements and much of our modern world would probably never have been developed.

✧ I don't understand why the creation of our technology rests on a particular value of the speed of light.

✦ Much of our technology rests on being able to transmit and receive electromagnetic waves, including radar, microwaves, radio and TV signals, and so on. The underlying theory that predicted electromagnetic waves was developed by James Clerk Maxwell over a century ago. None of this could have happened if the speed of light had been a million times greater.

✧ What does the value of the speed of light have to do with the development of electromagnetic theory?

✦ It is possible to understand magnetic fields as arising from a tiny relativistic correction to electric fields—a correction that becomes less significant the greater the speed of light becomes. Thus, it is quite possible that magnetism would never have been observed if the speed of light were a million times greater.

✧ So if the speed of light had not been measured and magnetism had never been observed, we obviously could never have developed a theory uniting electricity and magnetism, and we would never have predicted electromagnetic waves. Anything else that might have turned out differently?

✦ Yes. Einstein probably would never have developed his relativity theory either. The main clue that led Einstein to relativity was the behavior of electromagnetic waves, as described by Maxwell's equations. Without that clue, it is likely relativity would never have been discovered. Even if it had, its experimental verification probably would have been impossible, because the size of the predicted effects would have been a million million times smaller.

✧ No theories of electromagnetism or relativity, no modern technology— would there have been any good news in this scenario?

✦ The one big piece of good news would have been no nuclear weapons, which depend on Einstein's relativity, in particular, $E = mc^2$, where c is the speed of light.

✧ Suppose, despite everything we've said, relativity had still somehow been discovered and nuclear weapons built. Would they have been any different from the nuclear weapons that now exist?

✦ If the speed of light, c, were a million times greater, then c^2 would be a million million times greater. This means that a million million times more energy, E, would be produced from a given quantity of mass, m. A single one of today's nuclear weapons could destroy the world.

What if Planck's constant were much larger?

✧ I'm afraid I'm very uncertain about this topic.

✦ Right, you would be.

✧ I haven't a clue where I am.

✦ Actually, you might know where you are.

✧ Are we doing a "Who's on first" routine here? Let's start over again.

✦ Sure. Where are you right now, exactly?

✧ I'm sitting here right next to you, about 9 or 10 feet from the door.

✦ OK, since you said 9 *or* 10 feet, we can say your position is uncertain by one foot. With a tape measure you could, of course, narrow that uncertainty considerably. How about your speed; what might that be?

✧ Since I'm at rest, I'd have to say zero, but in the spirit of what you just said, I'd guess I'd say less than maybe one millimeter per second, if I take into account the possibility of small bodily movements.

✦ So, based on your rough estimates, we have a pair of uncertainties—one for your position and one for your speed. We might imagine reducing both of these uncertainties to arbitrarily small values by making more careful measurements. But as it turns out, there is in fact a minimum possible value for the product of the two uncertainties. We refer to this insight as the Heisenberg uncertainty principle.

✧ Is this something that I would ever notice in real life?

✦ No, because the minimum value of the product of the two uncertainties equals an incredibly tiny number—far smaller than you could ever encounter given more mundane sources of uncertainty.

✧ I'll bet that tiny value is Planck's constant.

✦ Actually, for your body it's Planck's constant divided by your mass—in general, objects having a very tiny mass, like say an electron, have a much larger value in the product of the two uncertainties.

✧ If I get your meaning, you are telling me that if somehow I managed to reduce the uncertainty in the position of an electron, I would find that its speed became more uncertain, so as to keep the product of the uncertainties above a certain value—sort of like reducing one dimension of an air bubble and causing the other dimension to grow.

✦ Right. One consequence of there being a minimum value of the position and velocity uncertainties is that an electron confined to a region the size of an atom would have a certain minimum speed, while an electron in the atomic nucleus—a region $100,000$ times smaller—would have to have a much larger minimum speed and hence much greater energy.

✧ Suppose we imagine a universe in which Planck's constant were enormously greater. What would things be like in that case?

✦ If Planck's constant (represented by the symbol $\hbar$—pronounced "h-bar") were increased by the ratio of your mass to that of an electron—10^{32}, or a billion trillion trillion—the consequences of the uncertainty principle would apply to you in the way it now applies to electrons.

✧ So if you put me in smaller and smaller rooms, my uncertainty in speed, and my speed itself, would grow larger and larger as the room size grew smaller. How could such a crazy thing happen?

✦ One way to understand the uncertainty principle is in terms of the wave nature of matter. By localizing an object more precisely, we are requiring that the accompanying wave "fit" inside the containing box, or that the object's "wavelength" be no larger than the box itself. The speed of an object is inversely proportional to its wavelength, so reducing the size of the box must increase the speed.

✧ I guess you could picture a tiger pacing more and more furiously the smaller its cage becomes.

The bartender at Planck's place has a lot of trouble getting drinks to stop in front of the patrons.

Chapter 12

Physics

What if you stood a pencil precisely on its point?

✧   If the pencil was slightly leaning in one direction, it would topple in that direction, but with no lean at all the pencil wouldn't know which way to fall, so I guess it would stay vertical indefinitely.

✦   Actually, even the most perfectly aligned pencil will eventually topple. If you place the pencil by hand, it's impossible not to give it a little push when letting it go, so let's suppose you place it using some kind of precision device. But even then, it would be impossible to prevent outside influences from toppling the pencil.

✧   How come?

✦   At any temperature there are unavoidable thermal vibrations present. Because of these vibrations, even the most precise method of release will give the pencil some small kick. Likewise, mechanical vibrations at the point of support will also disturb a pencil placed in the vertical position.

✧   How about if we imagine the pencil being placed at absolute zero temperature, where we can ignore thermal vibrations. Couldn't the pencil remain vertical in that case?

✦   Surprisingly, no. According to the uncertainty principle, if you precisely locate the top of the pencil so that it is exactly centered over the point of support, you unavoidably give the end of the pencil a large random kick. Likewise, the only way to avoid the random kick is to be less precise in locating the end of the pencil. In either case, the pencil will not remain vertical once you let it go.

✧   I thought the uncertainty principle applied only to atomic and subatomic systems, because the effects are too small to be noticeable for large-scale things like pencils.

✦   Normally, for large-scale objects, other disturbing influences are much more important than the limit imposed by the uncertainty principle. But here we are assuming that all those other influences have been eliminated. A pencil placed on end is an example of an unstable system. For such systems, even the most minute

146

deviation from the equilibrium position causes the object to move farther and farther away from that position.

✧   How long could a pencil stay on end before the tiny uncertainty implied by the uncertainty principle caused it to topple?

✦   That upper limit has been calculated to be around 5 seconds. But if you try to balance a pencil on its point, you will probably find that it topples in less than a second, because you won't be able to align it accurately enough with the vertical. Even if the alignment is off by only a hundredth of a degree, the pencil will take only about 3 seconds to topple over and hit the surface on which it rests.

✧   So is there no way to balance a pencil on its end indefinitely?

✦   Actually, there is a way. Take a yardstick and try balancing one end on your finger, while looking at its top end. If you continually move your finger appropriately, you can counter the tendency of the stick to topple. In principle you could do the same for a pencil, except it topples too quickly, so your reflexes probably aren't fast enough to move your finger. But if you have very fast reflexes, you might be able to do it if you put a clay ball at the top end to slow its toppling.

✧   Sounds like a nice thing to try at my next "bored meeting."

What if there were three kinds of electric charge?

✧   You could say there are three kinds of electric charge now: positive, negative, and neutral. But I guess you'd tell me that neutral is just the absence of the other two types.

✦   Right. It's also what you'd have if you mixed equal amounts of positive and negative charge, because the total effective charge would be zero.

✧   So I guess that's why we call the two types positive and negative rather than, say, chocolate and vanilla.

✦   Exactly. But suppose there were a third type—say, strawberry. How would we know it? For that matter, what makes you think there is no strawberry-flavor charge now?

✧   I don't like strawberry; let's call the third type pistachio. I guess its because no one has ever observed any pistachio charge.

✦   But we don't really directly observe any kind of electric charge. We just infer that a certain kind is present, based on its effect on other charges.

✧   Oh, you mean that positively charged objects attract negatively charged objects and repel other positively charged objects.

✦   Exactly. And for that reason we know that there must be at least two types of electric charge, not one. With only one type, everything would either attract everything else or repel everything else.

✧ Just like gravity.

✦ Right. That's why we say there is only one type of gravity (attraction), due to one type of gravitational "charge."

✧ But getting back to electrical charges, I guess the pistachio charge (if it existed) would have to be recognizable in some way, based on its different behavior.

✦ It need not be a "different" behavior; it could be the same behavior. Here's one possibility: Suppose objects with positive (chocolate), negative (vanilla), or pistachio charge each attracted objects having a different charge but repelled objects having the same charge.

✧ Let's get more specific. Suppose I had three objects: chocolate, vanilla, and pistachio ice-cream cones. What would I find if each had a different charge?

✦ Each cone brought near one of the others would be attracted to it. But if you took a bite out of one cone and brought the rest near your mouth, it would be repelled by some of the same kind of ice cream already in your mouth.

✧ Making it somewhat difficult to finish the cone, I guess. But how would this observation prove that the cones had three different charges?

✦ If cone C (the chocolate one) attracted both cone V (the vanilla one) and cone P (the pistachio one), cones V and P would both have to be a different flavor from cone C. In addition, since cone V and P attracted, they would be different flavors. So each of the three cones would be a different flavor, and there would have to be at least three possible flavors to explain the observations. To see if there were no more than three flavors, you could try a similar experiment with four cones having flavors you thought might be different.

✧ But suppose the observations had turned out differently? Is it possible that there might be three types of electric charge that would not be detectable in such experiments?

✦ Sure. Chocolate and vanilla might attract, but chocolate and pistachio might repel each other. In that case, how would we know that chocolate was any different from pistachio? Or maybe the pistachio type of charge would be much more difficult to isolate than the others and would hardly ever be found anywhere.

✧ Well, I do have a difficult time finding pistachio in the stores. So you mean there really could be a third type of electric charge that we just don't know about?

✦ No, probably not, because when we combine equal amounts of "chocolate" and "vanilla" charge we always get zero, which would be surprising (but not impossible) if there were some as yet unobserved third kind of charge.

What if height or other attributes were quantized?

✧ What exactly would it mean to say people's heights were quantized?

✦ People have heights that statistically follow a smooth bell-shaped curve—few people have heights much less than the average, and few have heights much greater. For any given range of heights, say from 70 to $70\frac{1}{8}$ inches, there are some people whose height lies in that range. But if your heights were quantized, they could only be exact multiples of some quantity. For example, if people's heights were quantized in units of one inch, no one would ever have a height other than an integral number of inches.

✧ Quantized height seems like a pretty far-fetched possibility. Your height changes continuously from birth, so it would have to advance in one-inch bursts in order to always be an exact number of inches. It would be as if people were constructed out of identical-size units, like the bricks of a building. As each "brick" got added, your height would change by one unit.

✦ Congratulations! You have discovered the key to quantization, namely, the existence of fundamental entities from which all larger objects are constructed. believe that charge is quantized because rge that is an integral multiple of some harge was considered to be e, the charge is believed to consist of quarks, whose of charge must be taken as $\frac{1}{3}e$. In any ltiples of $\frac{1}{3}e$, you'd never find an object

quarks now. What other properties be-

to the discovery of a long list of quan- given the cute names of "strangeness" matter outside particle accelerators. One to reconcile with our everyday-life ob- neasure of a body's rotational motion.

orhoods—I see plenty of examples of you mean about rotation. I assume that u are saying that I could spin with only f rotation could not vary continuously.[1] pin actually observable?

Planck's constant divided by 2π. Actually, the escribed above, and it depends on the type of category known as bosons have allowed spins

◆ Not for a person, because the size of the fundamental unit of angular momentum is so tiny. It would be like trying to observe the quantization of your height if the bricks from which you were made were only a trillionth of an inch high rather than an inch.

✧ Are there other interesting examples of things that are quantized?

◆ Time is normally considered to be continuous (not quantized), but some physicists have suggested that it too could be quantized. Let's call the hypothetical fundamental unit of time the chronon. If time were quantized, we would advance through time one chronon at a time. Life would be much like the individual still frames of a movie. It is only because the size of the hypothetical chronon is so tiny that we get the illusion of time advancing continuously.

✧ It seems to me that the idea of quantized time is clearly absurd. Suppose a particle discontinuously advanced from point A to point C in a time of one chronon. Then there would be no time at which it would be found at point B, located midway between A and C. How is that possible?

◆ This argument against the idea of quantized time was first advanced by the ancient Greek philosopher Zeno. Unfortunately, Zeno had an equally persuasive argument against the idea of time being continuous. His argument was that an object passing from A to C would have to pass through an infinite number of positions in a finite period of time, which also seemed absurd to him. Thus, Zeno reasoned that the passage of time had to be an illusion, since it could be neither quantized nor continuous.

✧ It sounds to me as if the possibility of quantized time will have to be settled by the physicists rather than the philosophers. Has anyone ever actually observed a chronon?

◆ The author of this book once thought he did, based on measurements of subatomic particle lifetimes, but the evidence turned out to be spurious.

✧ Too bad—it sounds like a neat Nobel Prize.

What if you could catch up to a light wave?

✧ I didn't know you couldn't.

◆ Actually, you could overtake light traveling in a transparent medium like water, where light travels slower than in a vacuum. But here we are considering a light wave moving through empty space.

✧ I just don't see why I shouldn't be able to catch up to it. I'm not talking

---

of 0, 1, 2, 3, . . . , units. Particles belonging to the category known as fermions (which include electrons and protons) must have an angular momentum of $\frac{1}{2}, \frac{3}{2}, \frac{5}{2}, \frac{7}{2}, \ldots$, units.

about practicalities, of course, like being able to build a fast rocket ship. Why should the laws of nature forbid me to do such a thing?

✦ That is a very good question. In fact it is the question that Albert Einstein asked himself while still a boy, and his conclusion that catching up to a light wave was truly impossible led directly to his theory of relativity.

✧ I hope I'm not going to have to be an Einstein to follow this. What was his reasoning?

✦ Think about any kind of wave, say a long succession of water waves. You can describe the waves in two ways—you can either look at the succession of crests and troughs at one frozen moment in time, as in a photograph, or you can stay at a fixed point and look at how the water height varies with time as waves pass that point. The mathematics of waves directly links these two descriptions.

✧ OK, so how would this apply to light waves? Incidentally, exactly what is "waving" in that case?

✦ A light wave does not consist of any material substance like water. Instead it is a variation in the strength of electric and magnetic fields. You could tell you were in the presence of a varying electric field, for example, because it would cause electrons in a wire, say, to bob up and down, creating an electric current— just as a cork that bobbed up and down in water could reveal the presence of a water wave even if you couldn't see the water directly.

✧ So if you did somehow (despite Einstein's belief) manage to catch up to a light wave, what would you observe?

✦ You would be like a surfer riding the crest of an ocean wave. You would observe a long train of ups and downs in the strength of the electric and magnetic fields in front of you and behind you. In this case, since the wave you were riding would not be moving from your perspective, you would see this variation of the fields stretching out in space but no corresponding variation of the waves in time. That is, you wouldn't feel yourself bobbing up and down like a cork. The mathematics linking the space and time variations, known as Maxwell's equations, would therefore be invalid.

✧ So?

✦ Einstein had enough faith in Maxwell's equations to say that if catching up to a light wave meant we had to abandon the equations, his choice was to say you couldn't catch up to a light wave.

✧ That sounds as if he preferred equations to reality.

✦ Not at all. Just that we have no experience with things like catching up to light waves, so who knows what the reality is? In fact, Einstein also gave an argument from experience. If it were possible to catch up to a light wave, then surely the speed of light would change when we observed light from sources that were moving toward or away from us. You would think, for example, that if the source were moving away at one percent the speed of light, the speed at which the light reached us should be reduced by one percent. But in fact, no such reduction is observed.

✧ So are you saying that the speed of light in vacuum is an absolute constant, unaffected by any motion of the source relative to us, and that this crazy fact has actually been observed?

◆ Yes on both counts, although the most compelling observations were not made at the time Einstein made his bold hypothesis. Nowadays, we can create a beam of particles traveling very close to the speed of light and observe the light that the particles emit while they move past. We observe a speed for the light that is unaffected by the emitting particles' motion.

What if energy were not conserved?

✧ I assume that when we speak of "conserving" energy, we are not using the phrase in the "good citizen" sense of using it efficiently and wisely.

◆ Exactly. We are here speaking of energy conservation in the scientific sense. Energy can exist in a wide variety of forms, including kinetic (energy of motion), potential (energy of position), heat, electrical, magnetic, and nuclear. When we say energy is conserved, we mean that a specific amount of one kind of energy is equivalent to a specific amount of another kind. Conservation of energy also implies that the total amount of all kinds of energy in a system that is isolated from its surroundings (a closed system) cannot change.

✧ Can you give a couple of examples?

◆ Sure. When water in a waterfall falls a certain distance, each pound of water can be said to lose a foot-pound of potential energy for every foot it falls, and its kinetic energy must increase by exactly the same amount to keep the total energy constant.

✧ What happens to all those foot-pounds of energy when the water hits bottom and "pounds the feet" of the falls?

◆ Nearly all the energy of the falling water is converted into heat, so the water at the bottom of the falls is slightly warmer than the water at the top. In fact, if energy is conserved, the increase in temperature will always be exactly the same when the water descends a given distance, assuming the water falls freely and doesn't hit any rocks on the way down.

✧ Suppose the water in the waterfall didn't fall freely but instead turned a waterwheel or turbine, and created electricity?

✦ In that case, a part of every foot-pound loss in potential energy of the falling water would show up as electrical energy and less would be converted to heat, so the water at the bottom of the falls would be warmed less. But the total energy in all its forms would remain unchanged—it would just be redistributed from one form to another.

✧ This business of energy conservation with fixed conversion rates from one form of energy to another sounds a lot like the conversion of money at fixed rates from one currency to another.

✦ That's a good analogy. If we have a specific amount of money to begin with, and convert it to a number of other currencies, the net amount of money remains unchanged. In practice, the analogy is flawed because currency traders charge commissions, and they have different exchange rates when they are buying or selling.

✧ So to pursue the analogy, what would it mean if energy were not conserved?

✦ If you started with $1,000, your funds at the end of a series of currency trades could be more or less than this amount—due to either continually fluctuating exchange rates or commission charges. Or, to get back to energy rather than money, the temperature rise in the waterfall would continually vary, the electrical energy produced from the falling water would vary, and the brightness of a 100-watt light bulb would also continually vary in an unpredictable way. It would be a highly confusing, unpredictable world.

✧ How else might the world appear if energy were not conserved?

✦ Another possibility would be getting energy for free—a perpetual motion machine. These imaginary devices require no fuel and are capable of providing unlimited energy.

✧ How can you be so sure they are not possible? Perhaps some clever inventor will devise one someday.

✦ Not likely. The failure of scientists and inventors to devise such machines or to find situations where energy is not conserved gives us more and more confidence that the conservation of energy is an absolute law of nature. If someone did come along with a machine that seemed to create energy from nothing, one possibility would be that it was tapping into an energy source that had until then escaped notice, though more likely it would be a fraud.

✧ So are you saying that the conservation of energy law has no loopholes whatsoever?

✦ Actually, there is one "loophole" provided by the uncertainty principle. But the violation allowed by the uncertainty principle is so small it would never be noticed in everyday life. According to this principle, it is possible to have "virtual" processes in which energy is not conserved, as long as the amount of "extra" or "missing" energy, multiplied by the time this discrepancy lasts, stays below a certain limit.

✧ This uncertainty principle business reminds me of a police policy of not bothering to go after embezzlers as long as the product of the missing funds and the time they are missing stays under a certain limit.

What if the oceans' heat could run an engine?

✧ Is there much heat in the oceans? Aren't ocean waters usually fairly chilly?

✦ Don't confuse heat and temperature. Any object that is above absolute zero temperature has internal heat energy, due to the kinetic energy of its randomly moving molecules. Temperature is a measure of the average value of this energy per molecule. The oceans, even though they are much colder than a hot cup of coffee, have enormously greater heat energy, because the number of molecules is so much greater.

✧ So why is there any problem in getting that heat energy out of the oceans to run an engine?

✦ Because heat flows only from hot objects to colder ones. In other words, the only way to use the heat in the oceans to run an engine would be to put the ocean water in some device containing a substance (say ice) that was appreciably colder than the oceans. In that case, a fraction of the heat from the oceans could be converted into useful work as the heat flowed "downhill" and melted the ice.

✧ OK, sounds good to me. Why couldn't we build an engine that way?

✦ Primarily because the energy needed to make the ice in the first place would be greater than the work the engine would produce.

✧ So are you saying that if we ever built an engine that extracted heat from the oceans, it would violate the law of conservation of energy?

✦ Not quite. You could imagine an engine that extracted heat from ocean water—in the process producing an amount of steam and ice that conserved energy exactly. The only problem is that the engine would have to make heat flow "uphill" (from cold water to the hotter environment) in order to work.

✧ But doesn't heat flow "uphill" in my refrigerator freezer, where heat is extracted from water that is converted to ice?

✦ Yes, but your freezer would not work if it weren't plugged in. An engine that produced work and ice cubes would be like a freezer that not only made ice for you but also generated mechanical work, all without being plugged in. Heat, like water, can be made to flow in the "uphill" direction only if it is pumped uphill—

which requires outside work. Left to itself, heat (or water) will always flow in the downhill direction.

✧ If such spontaneous "uphill" flows of heat were possible, what would the universe be like?

✦ If we assume that heat could flow spontaneously in both directions—from hot to cold objects and cold to hot ones—a variety of perpetual motion machines would be possible, one of which we described previously: extracting ocean heat to produce useful work, with ice cubes as a by-product.

✧ How do we know such perpetual motion machines aren't possible? Maybe some clever inventor could make one.

Thermal energy everywhere, but not a drop to extract . . . except in a cartoon.

✦ All our experience to date would imply the reverse. But since all the laws of nature are provisional (true until someone finds an exception), there is no way to prove mathematically that this is impossible. If you do build one, the Patent Office will want to see a working model!

✧ So getting back to the real universe, is there no realistic prospect for an engine that extracts heat from the oceans?

✦ Actually, nature has found a way to do just that. Large areas of ocean surface warmed by the sun cause rising columns of air that lead to the formation of hurricanes. A hurricane gets its energy from the warmed ocean surface.

✧ Might we be able somehow to imitate nature?

✦ Possibly. One potentially feasible way to build a power plant that uses the heat in the oceans relies on the temperature difference between surface and deeper waters. When heat flows spontaneously from the warm surface to the cold water below, useful mechanical work can be extracted.

What if you could unscramble an egg?

✧ I guess, in that case, it would be possible to put Humpty Dumpty together again. But why does nature not allow us to unscramble eggs, and what exactly do we mean by scrambling?

✦ It's all a matter of probabilities. Let's think of the yolk and the white part of the egg as each being made of entirely different kinds of molecules. Initially all the yolk molecules are in one region and all the white molecules in another. Let's use the word *scrambling* to exclude the chemical changes created by frying and to mean only moving the molecules around in a random way, that is, increasing their disorder, or entropy. When an egg is scrambled, it is extremely improbable that all the yolk molecules will wind up finding themselves located in the original location of the yolk—just as it is usually unlikely that a shuffled deck of cards will wind up with the deck in its original order.

✧ Why do you say that "usually" a shuffled deck is unlikely to be returned to its initial order?

✦ Sleight-of-hand artists can shuffle a deck "perfectly," meaning that the cards in the two halves of the deck are exactly interleaved. It can be shown that eight consecutive perfect shuffles will restore the deck to its original order.[2] But scrambling an egg is more analogous to shuffling a deck of cards *im*perfectly—

2. For example, consider card number 2 from the top of the deck. As you can easily verify, successive perfect shuffles move that card to the following sequence of positions: 2, 3, 5, 9, 17, 33, 14, 27, and finally back to 2 after the eighth shuffle—a result that occurs for a card in any position.

which injects an element of randomness into the process. In such a case, it is extremely improbable to have the original state restored by chance.

✧ But when you say it's extremely improbable, does that leave the door open to the possibility of its happening?

✦ The likelihood of its happening is so infinitesimal that, for all intents and purposes, we can say it's a practical impossibility. The reason is that the odds of unscrambling an egg (or unshuffling a deck) depend on the number of molecules in the egg (or cards in the deck).

✧ Why should the number of molecules or cards be the crucial factor?

✦ Consider a deck of only three cards initially in 1, 2, 3 order. There are only five other possible orders (as you can easily check), so a random shuffling has a one in six chance of giving back the original order. Thus, one in six shufflings will give you back the original order. But with ten cards in the deck, there are 3,628,800 possible orders. Even if you shuffled the cards once a second it would, on average, take over a month before you got back the original order by chance.

✧ Given the huge number of molecules in an egg, I can see how a random shuffling will never give back the original arrangement. What would the universe be like if you could unscramble eggs?

✦ In that case, presumably we would see order spontaneously arise from disorder in all kinds of ways. It would be much like watching a movie of the universe running backward. We would also presumably observe heat flow "uphill" from cold objects to hot ones.

✧ What is the connection between heat flowing uphill and unscrambling eggs?

✦ Suppose you heated one end of an iron bar and put the other end in ice. After you removed the bar from the heat and ice, the entire bar would eventually reach a uniform temperature. But we never see the reverse sequence. In other words, if the bar starts out at a uniform temperature and there are only random molecular movements, it is almost impossible for the vigorously moving molecules to wind up at one (hot) end and the sluggish ones to wind up at the other (cold) end. But it would happen in our "unscrambled egg" universe.

✧ So in the real universe can we never unscramble an egg?

✦ Actually, it might be possible, but you just can't do it by continued random scrambling. In practice, you can partly unscramble an egg by exploiting the difference in mass between yolk and white molecules. If you let a raw scrambled egg sit in a jar for about a week in your refrigerator, you will find that most of the lighter, yolk portion rises to the top—but you can forget about putting the yolk back together!

What if all the air in your room went into one corner?

✧   I guess I would suffocate. But such a thing is not possible, is it?—at least not without a huge vacuum pump in the corner that would suck the air in.

✦   Some events, like the one we are considering, are theoretically possible, but the probability of their actually occurring is so small that for all intents and purposes they are impossibilities.

✧   I don't understand how it is even theoretically possible for it to happen.

✦   Let's imagine that the corner of your room occupies a volume $1/10$ that of the room itself, and ask how likely it is that a given molecule due to its random motion will be in that tenth of the room.

✧   Clearly, if the motion is random, the probability of any one molecule being in the corner is $1/10$, or 10 percent, at any given time.

✦   Right. Suppose we follow the motion of two molecules and ask about the probability that both of them will be in the corner. Assuming the two molecules move independently of each other and each has a 10 percent chance of being in the corner, then the chance of both being in the corner is 10 percent of 10 percent, or one percent.

✧   So if we were to extend this, say, to 100 molecules, the chance of finding all 100 in the corner would be $1/10^{100}$. That sounds like a pretty small number.

✦   Yes the number 10^{100}, written as a one followed by 100 zeros, is called a googol, so the chances of finding all 100 molecules in one corner of the room would be one in a googol.

✧   Suppose we actually had a room in which there were only 100 molecules. How long would we have to wait for all 100 to be in one corner?

✦   At room temperature, the speed of an air molecule is in the neighborhood of a few kilometers per second. In a typical-size room containing only 100 molecules, a single molecule might move back and forth between a corner and the rest of the room perhaps once every thousandth of a second, given its speed. That means we would have to wait a thousandth of a googol, which we could call a milligoogol (10^{97}), seconds for the 100 molecules all to be in one corner.

✧   I once dated a girl named Millie Googol. How long a time is that?

✦   The universe is believed to be about 10 billion years, or roughly 10^{17}, seconds old. In other words, we would have to wait 10^{80} times the age of the universe for all 100 molecules to wind up in the corner. That number, 10^{80}, also happens to be roughly the number of protons in the entire universe.

✧ So far we have only been considering the odds of 100 molecules randomly moving into the corner. I imagine there are quite a few more molecules actually present.

◆ Yes. In a typical-size room there are approximately 10^{27} air molecules. The probability of all of them being in the corner at any one moment due to their random motions would be one over ten raised to the 10^{27} power. No comparison can easily convey the smallness of this number. It is comparable to the chance of a statue spontaneously jumping into the air.

✧ How did we get from air molecules to jumping statues?

◆ The molecules in the ground below the base of the statue also have random motions. If at one moment most of them happened to be moving upward, they would propel the statue upward.

What if you sawed the ends off a magnet?

✧ I'm tempted to say that you would have liberated the poles, but I don't want to be accused of making any Polish jokes.

◆ Actually, of course, after sawing the end off a magnet, you would find that both pieces had new poles. In fact, no matter how many pieces you sawed off, each one would have north and south poles at opposite ends.

✧ Where do all the poles come from?

◆ Now you're trying to get me to make a Polish joke. You could say that the magnetic poles were there all along. Think of the iron atoms in a bar magnet as being miniature magnets. An appreciable fraction of these atomic magnets are aligned end-to-end along the length of the magnet. So when you saw the magnet at any point, one side of the cut has north poles on the surface and the other side has south poles.

✧ But why couldn't you get separate north and south poles by sawing one of the atomic magnets in half?

◆ Have you ever made an electromagnet by connecting a battery to a wire wound around an iron nail? At the atomic level all magnets are electromagnets. That is, atoms produce their magnetism only because of the electric currents created by orbiting and spinning electrons. The concept of north and south poles may be a useful way of talking about large-scale magnets, but there is no such thing at the atomic level, or at least that's the standard theory.

✧ Do I detect that you are leaving the door open a crack on the question of whether unpaired north or south magnetic poles might exist?

✦ Yes, as a matter of fact. It is not possible to say with certainty that something does not exist. In the case of magnetic poles, P.A.M. Dirac, a well-respected physicist, has claimed on theoretical grounds that such a thing, called a monopole, actually should exist. People have even searched for these magnetic monopoles.

✧ How would you go about such a search? I have visions of people at a magnet factory going around and sawing the ends off all the magnets looking for monopoles.

✦ No, monopoles, if they exist, are probably bombarding Earth from space in the form of invisible subatomic particles.

✧ What would be the distinguishing characteristics, or the "signature," of a monopole that would allow you to say you had found one?

✦ Suppose you were to drop a bar magnet through a coil of wire, to which you had attached a sensitive current meter. You would find that as the north pole of the magnet fell through the plane of the coil, an electric current would circulate around the coil one way, and then a moment later, as the south pole fell through, the current would circulate in the opposite direction. Such induced currents are the basis of much of our technology, including electric generators, telephones, microphones, and transformers.

✧ Based on the opposite current flows caused by the a north and south poles of a magnet, I assume we could set up a coil of the type you mentioned to detect magnetic monopoles. Presumably, if we found an induced current that flowed only in one direction, that would be a sign that a magnetic monopole had passed through the plane of the coil.

✦ That is exactly what is done. Based on this search method, the grand total of magnetic monopoles found to date has been exactly one. But that one "event" occurred in the middle of the night, so there may be alternative explanations—such as a graduate assistant's pulling a hoax. It is now especially difficult to put any credence in that one event, since none has ever been seen since.

✧ I guess that would be similar to hearing your dog speak once in its entire life—you would probably think it was a delusion. Although, then again, it might depend on what the dog said.

What if everything were reflected into its mirror image?

✧ I used to have trouble keeping track of which hand was my right as a kid, so I had to label them R and L. Couldn't we use that simple labeling scheme to tell if such a mirror reversal took place? My L- and R-labeled hands would switch places, so wouldn't it be easy to tell that a reflection had occurred?

✦ Not really, because how would you know they had switched places? If you had a photograph of how you looked before the switch, that photograph, along with everything else in the universe, would be reversed.

✧ I guess it wouldn't do any good to define the left side of my body as the side my heart is on, because my heart would also get switched. So are you saying it would be impossible to tell if the switch occurred?

✦ Until 1956 that's what most scientists would have said. Before we describe what happened in that year, let's think about how you would explain the meaning of right and left to an extraterrestrial scientist with whom you had established radio communications.

✧ Couldn't I just transmit a TV picture of me with my right and left hands labeled?

✦ Actually, that wouldn't work, because to translate the digital sequence of zeros and ones of your message into a picture, the alien would have to decide whether each scan line of the image started on the left or the right. The two choices would result in mirror image pictures of you.

✧ Obviously, if I could send the alien a physical object, I could easily explain the meaning of right and left, but I guess that's "cheating."

✦ Right. On the same basis, we'll also call it cheating if the radio signal you sent the alien were circularly polarized. In this case, if he were to look toward the direction from which the signal came, he would see the electric and magnetic fields rotating either clockwise or counterclockwise. If you transmitted the clockwise wave, you could then explain that your left hand was the one whose fingers curled clockwise when the thumb pointed toward him, or her—or it?

✧ Well, maybe they would all be robots. Anyway, couldn't we use some of the basic laws of magnetism to describe the meaning of left and right, or clockwise and counterclockwise? For example, couldn't we tell the alien to wind a wire around an iron bar, pass an electric current through the wire, and see which end of the bar became the north pole of the electromagnet? This would unambiguously tell whether the wire had been wound clockwise or counterclockwise around the bar.

✦ Unfortunately, explaining to the alien how to tell which end of a magnet is its north pole would be no easier than explaining left and right. You might think that on Earth we do have an absolute way to define the north pole of a magnet with reference to Earth's north pole, but that is not so. Over the course of Earth's history, its north and south magnetic poles have switched places. So if we defined the north pole of a compass needle in terms of which end pointed north, that definition would reverse when Earth's magnetic field reversed.

✧ So what happened in 1956 to change our ability to tell the alien the meaning of left and right?

✦ An experiment was done using a sample of radioactive cobalt, whose nuclei we can think of as miniature bar magnets. When radioactive cobalt nuclei decay, they emit electrons, which would be expected to be emitted with equal likelihood in any direction. But when the cobalt nuclei were all aligned by placing them in a strong magnetic field, the electrons were emitted preferentially in the direction of the magnetic field along which the nuclear "bar magnets" were aligned.

✧ So what does this experiment have to do with left and right and mirror reversals?

✦ Picture the experiment placed in front of a mirror. If the mirror-image experiment were actually built (with the electromagnet coils wound the opposite way), the direction of the magnetic field and that of the emitted electrons would both reverse. This experiment therefore would give us an absolute way to explain to the alien which end of its electromagnet was north—regardless of whether he built the original or mirror-image version. We could just tell the alien that the north pole was the direction in which electrons from aligned radioactive cobalt nuclei tended to be emitted. With an absolute definition of magnetic north, he could then define left and right, as we previously described.

✧ No more catches?

✦ Actually, there is just one. We would need to be sure that our alien friend were made of matter, not antimatter. It is believed that the same experiment done with anticobalt nuclei would have exactly the opposite result.

✧ Considering that matter and antimatter go "poof" when they meet, we might want to be wary of any aliens extending their left hand in greetings!

What if chemical reactions required a critical mass?

✧ The concept of critical mass originally comes from nuclear weapons and nuclear power, right?

✦ Exactly. Certain elements such as uranium or plutonium are fissionable, that is, they have isotopes whose atomic nuclei can be split, or fissioned, spontaneously. A nuclear chain reaction can occur if more than a certain critical mass of such fissionable material is gathered at one place.

✧ With the result being a nuclear explosion?

✦ It would be a nuclear explosion only if the reaction took place so fast that the material would not be blown apart from the heat produced, and kept going. This can't happen in a nuclear reactor. If a nuclear reactor "goes critical" as a result of an accident, the fuel will be blown apart before the reaction proceeds far enough for a nuclear explosion to occur. A lot of heat and radioactivity can be released, and possibly there could even be a "meltdown," but no nuclear explosion.

✧   What is the basic idea of a chain reaction?

✦   When a uranium nucleus fissions, or splits, it also emits neutrons—typically two of them. If the two neutrons emitted in any one fission are absorbed by two other uranium nuclei and they fission as a result, the number of fissions will double each "generation." Since each generation lasts only about a microsecond (a millionth of a second), the number of fissions increases by an enormous factor in a very short time.

✧   Just like if you gave two friends a cold and they each infected two friends, there would be an epidemic in short order. But why does there have to be a critical mass for a chain reaction? Why doesn't it always happen?

✦   Because the neutrons emitted by one fission travel on the average a few inches before being absorbed by other uranium nuclei. If the mass of uranium is too small, too many neutrons escape through the surface of the material without causing further fissions. Think of a bunch of people in an auditorium who are spreading rumors by walking over to whisper in the ears of people located ten feet away in a random direction. No rumors would spread if everyone were inside a circle of diameter less than ten feet.

✧   What is the critical number of surviving neutrons you need each generation to get a chain reaction?

✦   The critical number is one—just as in the case of rumor-spreading. For a mass even slightly larger than the critical mass, more than one neutron survives from each fission to create a new fission, and the number of fissions grows geometrically or exponentially.[3]

✧   How big is the critical mass needed to make a nuclear bomb?

✦   For plutonium it's around 7 kilograms, or 16 pounds. To make the bomb explode, all you need to do is bring two pieces whose total mass exceeds 7 kilograms together very quickly using conventional explosives.

✧   What would happen if chemical reactions required a critical mass, like nuclear reactions?

✦   You would want your gas tank to emulate a nuclear reactor rather than a nuclear bomb, so you would need to be extremely careful in filling your gas tank, lest the amount of gas exceed the critical mass! Actually, gas tanks and engines

---

3. For example, if on the average 0.93 neutrons escape through the surface, leaving 1.07 to react, the number of fissions would double in ten generations. In a thousand generations (lasting a mere thousandth of a second), the number of neutrons would increase by a factor of 2^{100}, which is around 10^{30} (a million trillion trillion).

would need to be redesigned so as to have a separate compartment where the amount of gas could be precisely controlled and be kept close to, but not greater than the critical mass.

✧ How come in the real world there is a critical mass for nuclear reactions but not for chemical reactions?

✦ Because with chemical reactions the heat energy liberated each time a molecule of fuel burns makes other molecules in the immediate vicinity move faster and start to burn. So it doesn't matter how much of the material is present. But for a nuclear reaction, the neutrons often travel so far before causing another fission that too many leave the material without causing a chain reaction if less than critical mass is present.

What if nuclear batteries existed?

✧ Didn't a couple of chemists claim to have invented exactly that a few years ago?

✦ Had it proved correct, the discovery of "cold fusion" by Stanley Pons and Martin Fleischmann would indeed have created the possibility of nuclear batteries, or at least heat energy from nuclear reactions in a battery-size cell.

✧ What exactly did the two researchers find?

✦ The two chemists made a fuel cell by placing platinum and palladium electrodes in heavy water (water in which the hydrogen nuclei have an extra neutron). They reported that a large amount of heat energy was created when an electric current passed through the cell. But actually their claim was based on their finding that slightly more heat was given off than they thought should have occurred, given the amount of electricity consumed by the cell.

✧ That sounds like having your company project an annual profit of $1 million based on an estimated income of $1,000 million and estimated expenditures of $999 million. I can understand why people might be a little skeptical about this kind of result. Were other scientists able to repeat their findings?

✦ Initially, there were some scattered confirmations, along with quite a few negative findings. The failure of Pons and Fleischmann to release the details of their experiment made it difficult initially to duplicate it elsewhere. But the consensus now is one of extreme skepticism toward cold fusion.

✧ What is the basis of that skepticism?

✦ From the outset, most physicists had difficulty believing that fusion reactions like the ones that power the Sun could be initiated at room temperature. To bring a pair of hydrogen nuclei close enough together to fuse, you need to overcome their electrical repulsion. The standard way to achieve such close contact is by heating the hydrogen to millions of degrees, so that nuclei collide with enough force to overcome their repulsion.[4] In fact, heating is considered the *only* plausible way to cause fusion—just as if you saw two "fused," scrunched up cars at an intersection, you could plausibly infer that a violent collision had taken place. Physicists were also skeptical about cold fusion because the reaction supposedly took place without the release of any nuclear radiation.

✧ If cold fusion had been real, what impact might it have had?

✦ It might have led to an essentially inexhaustible supply of cheap energy. Heavy water constitutes one part in 600 of water in the oceans, so we would have thousands of years of fuel available at very low cost. If cold fusion could really have produced energy without the radioactivity that accompanies nuclear fission reactions, it would have been a great boon to society.

✧ Speaking of nuclear fission, how about the possibility of a battery based on that process? Unlike nuclear fusion, you don't need temperatures of millions of degrees to initiate a fission reaction. And don't nuclear-powered "batteries" already exist in some satellites?

✦ Unfortunately, the smallest nuclear reactors, used to power submarines, are still quite sizable. The only way we know of to get energy out of a fission reactor is through the heat produced by the reaction. The heat converts water into steam, which powers a turbine that can be used to create electrical energy. In addition, you usually need to shield the reactor to prevent the escape of radiation. All these processes require a massive amount of equipment surrounding the reactor core. It is always hazardous to predict future technological developments, but it seems difficult to conceive of all these processes occurring in battery-size cells. The nuclear power supplies of some satellites are based on the heat generated from radioactive decay, and are not true nuclear reactors.

✧ Are there any other possibilities for a nuclear battery?

✦ If you want a really far-out example, we could imagine an antimatter battery. When antimatter reacts with ordinary matter, the result can be a complete conversion of matter into energy.

✧ Where do you find antimatter?

✦ Some antimatter may be in your room right now—the result of cosmic rays colliding with atoms of air and producing positrons (antielectrons). But don't expect to find any supplies of the stuff just lying around. At present, we can only produce minute amounts of antimatter from particle collisions in high-energy accelerators. In fact, the amount of energy needed to produce antimatter greatly exceeds the amount of energy it delivers when it reacts. For the foreseeable future, antimatter energy storage is strictly for the starship *Enterprise*.

---

4. Actually, it is not necessary for the nuclei to make contact in order to fuse, because of a process known as quantum mechanical tunneling.

Chapter 13

Sun and stars

What if the solar system had two suns?

✧   Are double suns a realistic possibility?

✦   Scientists have discovered that many stars in the sky, perhaps even most, are not single objects, as they appear to the naked eye, but two or more stars fairly close together. Planets around such stars might orbit two suns. But there are a couple of catches.

✧   Like what?

✦   We don't know what fraction of stars have planetary systems, but it is possible that planets are a "substitute" for another sun.

✧   So in that case, stars might have either a companion star or planets, but not both. What is the other catch?

✦   The orbits of planets around a single massive sun can be highly stable, and last indefinitely. But orbits about a pair of suns may not always be stable; that is, a planet might eventually crash into one of the suns, or fly out of the system.

✧   Are there any ways to get stable planetary orbits in a two-sun system?

✦   The two most favorable possibilities are at opposite extremes, where either the two suns are very far apart and the planets basically orbit one of them, or the two suns are very close together compared with the sizes of the planetary orbits.

✧   Presumably, in the second case, an observer on a planet during daytime would see two nearby suns in the sky orbiting each other.

✦   Right. Typically, pairs of stars orbit each other with a period measured in days, so even though you wouldn't see the stars actually moving in their orbits, you would definitely see them change their relative positions each day. On some days one sun might even eclipse the other, possibly leading to drastic changes in brightness if the cooler one passed in front of the hotter one.

✧   It would seem weird to have two sunrises and two sunsets a day. Or could there be a way to have one sun rise during the "day" and the dimmer of the two suns rise at "night"?

✦ Yes, in fact that is quite possible. The situation you describe would occur when the planets basically orbited one of the two suns, and the size of the orbits was much less than the distance between suns.

✧ I'm not sure I get the picture.

✦ Put two quarters far apart on a table representing the suns, and put a dime near one of them representing a planet. The dime orbits the closer quarter with a period we'll call one year. During the part of the year the dime is between the quarters, both suns lie in opposite directions as seen from the dime, so the near one would be visible during the day and the distant one at night. Of course, if the distant one were very far away, it might appear more like a very bright star than a sun.

✧ I feel deprived living in a solar system with only one sun.

✦ Actually, maybe you need have no reason to feel deprived. Some scientists have suggested that our Sun may actually have a very faint companion star orbiting it, called Nemesis. The basis of the suggestion is that species seem to have become extinct in large numbers during certain epochs in Earth's history, and these so-called mass extinctions seem to have occurred on a regular basis, which could be explained by Nemesis. The idea is that Nemesis, during each orbit around the Sun, might shake loose some comets from the distant outer reaches of the solar system. Some of the comets might collide with Earth and cause many species to become extinct. If the size of Nemesis's orbit were large enough, it would appear quite faint, and we wouldn't be able to distinguish it from any other star.

✧ But surely astronomers could observe that Nemesis was actually orbiting the Sun, and was much closer than all the other stars we see.

✦ Astronomers could, if they could pick it out among the billions of stars too faint to be seen by the naked eye. But without knowing precisely where to look, they would have to start measuring the distances to all of them, which is beyond our present capability.

✧ So I guess for now old Sol's companion will remain his secret.

What if the Sun's temperature were 600 degrees Kelvin?

✧ Why did you call me Kelvin?

✦ Not you—it's a unit of temperature. Kelvin, or absolute, temperatures are measured relative to the coldest possible temperature (−273 degrees Celsius).

So by definition, a temperature on the Kelvin scale is 273 degrees higher than its Celsius value.

✧ Six hundred degrees Kelvin would be 327 degrees Celsius. That sounds like a pretty cool Sun. How hot is the Sun actually, and is there any reason why it has that temperature?

✦ Its core temperature is calculated to be about 15 million degrees Kelvin, but its surface is measured to be only about 6,000 degrees. The Sun, being a ball of gas, collapses under its own gravity until the pressure from the heat generated at its core just counters the inward crush of gravity. For a star the mass of our Sun, that balance is struck when the surface temperature is 6,000 degrees Kelvin.

✧ How can we measure the Sun's temperature?

✦ An object that is hot enough to glow emits light having a color characteristic of its temperature. The hotter the object, the more the color shifts toward the blue (short wavelength) end of the spectrum. Think of how the color of an electric stove's heating element changes as it heats up, and you'll get the idea.

✧ Let's see, when I look at a heating element after it turns on, I see it first turn red, then orange, and then yellow. From what you said, a much hotter object would eventually appear blue.

✦ That's right. Those stars that appear blue are, in fact, very much hotter than our Sun. Astronomers measure not only the average color of the Sun but also the amount of light present at each wavelength—the spectrum. The Sun's spectrum is found to be approximately the same as that of an ideal blackbody.

✧ The astronomer who decided the Sun was a blackbody must have been color-blind—excuse me—I mean chromatically disadvantaged.

✦ Blackbodies have that name because they absorb all light that falls on them. But they actually appear black only at very low temperatures.

✧ So based on what you said about the connection between temperature and color, if the Sun were only 600 degrees rather than 6,000 degrees Kelvin, I assume its color would be much redder.

✦ Yes, and if life were to evolve around such a sun, animals would probably have infrared vision, and plants would rely on a different version of photo-synthesis.

✧ Any other problems for the evolution of life?

✦ Yes. The total amount of light emitted by the cooler sun would be far less than it is now, assuming we had a sun of the same size. In fact, the total energy output of a sun the same size as ours would be proportional to the fourth power of the absolute temperature.

✧ I assume that means if the temperature were 10 times less, then the energy emitted would be 10,000 (10^4) times less. That sounds like bad news for life on any planets around such a sun.

✦ Yes, and in order to have the same temperature as Earth, the planet would need to be a lot closer to its sun than Earth is to our Sun.

✧ Just like when a campfire dies down, everyone needs to move closer to the fire to stay warm. How much closer would the planet need to be?

✦ Picture two imaginary spheres centered on our Sun—the first one having a radius equal to Earth's distance from the Sun, and the second 100 times smaller in radius (which makes its surface area 10,000 times smaller). The same total amount of sunlight passes through both spheres, so the amount of sunlight passing through each square meter of the small sphere is 10,000 times greater.

✧ I see. So a 10,000-times dimmer sun would, at the 100-times closer distance, produce the same warmth as the actual Sun does at Earth's location, since the two factors of 10,000 just offset each other. Could life exist on a planet around such a dim sun?

✦ Possibly, but it's not too likely. Because of its nearness to its sun, the planet would experience tidal forces a million times greater than those on Earth. In relatively short order those tidal forces might reduce the planet's rotation, leaving one side blazing hot as it faced the sun and the other side in freezing darkness.

✧ How would our Sun actually appear in the sky under the conditions we are imagining?

✦ If we were 100 times closer to the Sun than we are now, the Sun's apparent diameter in the sky would be 100 times greater. The Sun would literally fill more than a quarter of the sky, and it would appear motionless in the sky halfway above the horizon as seen from the only place on the planet life could exist—the transition zone between the bright and dark sides.

✧ Sounds like something Van Gogh might paint.

What if the sun became a supernova?

✧ OK, what's a supernova?

✦ It's a once-in-a-lifetime explosion of a star that literally blows itself to bits. During a supernova explosion, a star emits so much light that it shines as brightly as all the hundreds of billions of stars in the galaxy.

✧ Sort of a last fling, perhaps. Where do stars get their energy from anyway?

✦ During the bulk of their "lives," stars generate energy when the nuclei of hydrogen atoms in their core collide and fuse together to produce helium nuclei.

Energy is produced because the helium is a little lighter than the hydrogen, and the loss of mass creates the energy, based on the relation $E = mc^2$.

✧ What happens to a star when it finishes fusing the hydrogen in its core?

✦ At that point the star collapses, because there is no longer any outward pressure from the heat generated at the core to counter the inward force of gravity. The gravitational collapse squashes the helium until it is hot enough to fuse and produce carbon. Eventually the helium is used up, and the star begins to collapse once again. After a whole series of such collapses and ignitions, the star finally reaches a crisis.

✧ I take it the star is approaching the supernova stage.

✦ Yes. After the star fuses a series of progressively heavier elements and produces iron at its core, it has reached the point where it can no longer stave off further collapses by fusing the "ash" left over from the previous fusion. For elements heavier than iron, fusion doesn't produce energy but rather consumes it.

✧ So how does the star resolve its crisis?

✦ The interior part of the star collapses into an extremely dense, hot object, and the outer, hydrogen-rich layers ignite when they collapse onto this superhot core. Then the star explodes.

✧ Will this eventually happen to old Sol?

✦ Actually, no. Only stars having about 40 percent more mass than our Sun have enough gravity to reach the point where iron is "cooked up" in their cores. Lighter stars like our Sun eventually go out with a whimper rather than a bang.

✧ Well, that's reassuring. You had me a little concerned that our Sun might blow up.

✦ Although our Sun won't blow up, you should be thankful that its ancestors did. All the elements on Earth heavier than iron—including the gold in your teeth—were created in the supernova explosions of stars whose remains found their way into the cloud of dust and gas that collapsed to form our solar system.

✧ If our Sun will be spared the dramatic final ending of a supernova, what will eventually happen to it after most of the hydrogen in its core has been fused?

✦ At that point the Sun will become a red giant. It will become so large that it will balloon out past half the radius of the orbit of Earth, and life on Earth will, of course, become impossible. But before you get too concerned, let me assure you that this won't happen for another 5 billion years or so.

✧ Is that the end of the line, as far as the Sun is concerned?

✦ After becoming a red giant, the Sun will settle down to become a white dwarf—an extremely compact star—and eventually it will gradually lose its heat, becoming a black dwarf—effectively a burnt-out cinder—its final state.

✧ How do "they" know all this stuff?

✦ Even though stars live far too long for astronomers to see any one star go through these various phases, they can test their models of the life cycle of stars by seeing if they find the predicted numbers of stars at different stages. To make an analogy, suppose a Martian had a theory of how humans evolve during the course of their lives. Let's assume the Martian was only able to visit Earth only for one day to test the theory. In principle, the Martian could check the theory by seeing how many people there were of each size and shape and comparing those observations with the theory's predicted numbers.

What if the stars in the sky were mostly planets?

✧ Since planets look more or less like stars, except for their lack of twinkle, I don't see why it would make any difference at all.

✦ Actually, it could make all the difference in the world. Planets, because they orbit the Sun, appear in slightly different positions from night to night against the background stars. If our solar system had a thousand planets and if there were very few true stars visible in the night sky, the pattern seen in the night sky would be a constantly shifting one.

✧ It looks like a constantly shifting one to me anyway. I can never see all those crabs, lions, and bears that are supposed to be there.

✦ That's not surprising, but people in ancient times, with clearer skies and without electric lights, were much more aware of the night sky than we are today. The regular patterns seen night after night made it possible for anyone with even a passing acquaintance with the sky to easily pick out the planets, which, being "wanderers," were the exceptions to the regular pattern.

✧ Why is it important that there are just a handful of wanderers among the stars?

✦ Because that's what motivated ancient peoples to observe and explain planetary motions. This probably would not have been the case if most of the sky had been filled with planets, and hence there had been no regular background star pattern against which to track the motion of individual planets. With the pattern constantly shifting, you might not even be able to tell which planet was which on successive nights.

✧ OK, granted the science of astronomy might not have "gotten off the ground." What else would have followed from a planet-filled night sky?

✦ Astronomy is the oldest of the sciences. Many discoveries in astronomy led directly to discoveries in physical science, such as Newton's law of gravitation. In fact, it might be that without first having seen nature's regularities in the night sky, humans would never have anticipated regularities elsewhere. Science in general might have made much more limited progress under a night sky filled mostly with planets.

✧ Is it realistic to imagine a solar system with thousands of planets?

✦ That depends on exactly how the solar system was originally formed. The standard picture is that the solar system started with a swirling mass of dust and gas. Colliding dust grains stuck together to form small bodies of up to boulder size, and the rotation of the whole system caused the coalescing bodies to form a flattened disk.

◇ So where did the planets come from?

◆ The many small bodies would have tended to coalesce further due to their mutual gravitational attraction, until eventually something like nine planets would have formed—at least that's what computer simulations show. On the other hand, had the original cloud been much heavier and swirling with a much greater rotational speed, many more than nine planets might have formed. Intelligent beings in such a solar system, unable to distinguish stars from planets, might have remained forever in the dark as to the nature of their universe.

◇ But why would an extraterrestrial civilization's scientific progress have to take the same path as ours? Why couldn't it—or ours, for that matter—have started with, say, biology instead of astronomy and physics?

◆ For the science of biology to progress from the descriptive stage to the deeper stage it has now reached, an understanding of the underlying physics was essential.

◇ Spoken like a true physicist.

What if the stars shone once in a thousand years?

◇ Isn't there a line in a poem by Emerson along those lines?

◆ Yes. The line goes: "If the stars should appear one night in a thousand years, how would men believe and adore, and preserve for many generations the remembrance of the city of God!" Isaac Asimov used this idea in his classic story "Nightfall."

◇ How could it happen that the stars would be visible only once in a thousand years?

◆ Asimov's story takes place on a planet in orbit about six suns, so invariably at least one is always in the sky. But every 2,500 years, an eclipse creates total darkness. The astronomers in the story have just discovered the means to predict such an eclipse, and their world awaits the impending darkness with great trepidation.

◇ I imagine people would be in great fear of darkness, never having seen it. One could even imagine people believing that the onset of darkness would bring the end of their world.

◆ That's exactly the case in Asimov's story, where the prophecy of religious cultists is that many will go mad, burn down the civilization, and start a new cycle over again every 2,500 years—a prophesy disbelieved by the scientists preparing to witness the eclipse. The cultists are also the only ones who anticipate the appearance of things called "stars" at the onset of darkness, as they have carried this knowledge forward in their Book of Revelations.

✧ What do you think people's actual reactions would be in a scenario such as Asimov describes?

✦ Very likely the recording of the event would take on profound mystical and religious meanings, particularly if a record of it survived for as long as 2,500 years. Even if a strictly factual record were made of the event, people in later generations would possibly give it mystical significance, never having witnessed such a thing. In a world in which it was always light, people might well become highly irrational when the darkness descended—particularly if there were historical evidence of worldwide catastrophe during these rare periods of darkness. At the conclusion of Asimov's story, people burn down their cities to ward off the unbearable darkness, and the prophecy of destruction becomes self-fulfilling.

✧ How about people's reactions to the sight of the stars in such a world?

✦ A world that had never seen the stars would have a much more limited conception of the size and scope of the universe. Of course, just being able to see the stars would be no guarantee that people would understand what they were seeing. In our own world, ancient peoples used to think of the stars as simply bright dots on some rotating "celestial sphere." Once people realized that the stars are actually distant suns at very great distances, our picture of our place in the cosmos was profoundly transformed.

✧ It is hard to imagine what it would be like to see something as majestic as a starry night once in a thousand years, and the feelings of awe such a sight would inspire, especially if this were part of one's religious beliefs.

✦ If you want a close analogy, suppose an eclipse of the Sun occurred once in a thousand years (which could happen if a lunar orbit lasted close to a year). In that case, one could well imagine the event taking on deep religious significance and even being written about in sacred texts. Even this analogy, however, doesn't quite indicate how awe-inspiring an event we would be dealing with, since solar eclipses are known events in our world. Having the stars appear once in a thousand years might perhaps be comparable to witnessing the stars in our world spontaneously arranging themselves to spell out the name of God.

Chapter 14

Space

What if the big bang wasn't so big?

✧   Are you suggesting a *little* bang theory for the creation of the universe? Or maybe that our universe was created in a pop or a ping rather than a bang? How can you possibly quantify the size of the "bang" that created the universe?

✦   The same way as in any explosion. The size of the bang can be quantified by its results—the speed with which the matter in the universe flies apart in its aftermath.

✧   So if the big bang hadn't been quite as big, the galaxies would be flying apart at somewhat lower speeds than now is the case.

✦   Right. And we would be able to observe that the distant galaxies showed less of a redshift, indicating a lower speed of recession than is now the case.

✧   OK, what's a redshift?

✦   That means the light we observe when looking at objects is shifted to longer wavelengths. It happens to any type of waves when their source is moving away, since then fewer waves reach you each second. Did you ever notice how the pitch of an ambulance or police siren appears to change as it passes?

✧   Right. It's more high-pitched when approaching and more low-pitched when receding. So I guess you'd say an approaching galaxy was blueshifted, since blue is at the high-frequency end of the visible spectrum. But why would anyone but astronomers care about whether distant galaxies had less of a red-shift than now?

✦   Because in that case the galaxies would not have had enough speed to escape one another's gravity, and the universe would be "closed."

✧   That's ridiculous. How can the universe not be closed? There's no way to get outside it, is there?

✦   When we say the universe is open or closed, all we are referring to is whether matter has enough speed to escape other matter's gravity.

✧ Like when something is shot upward fast enough from the surface of Earth it can escape Earth's gravity?

✦ Right. Also, since gravity affects the path of light rays, in a closed universe a

Having decided to start things off with a bang, God selects its size.

light ray traveling in what seems to be a straight line would in theory eventually come back to its point of origin, if it weren't absorbed.

✧   I assume that's sort of like the way someone traveling on Earth should eventually come back to their starting point if they go in a "straight line" far enough.

✦   Yes, with the important difference that it is possible to get off the two-dimensional surface of Earth and see that our "straight line" on Earth is really a great circle. But we know of no way to get out of the three-dimensional space (or actually the four-dimensional space-time) of our universe.

✧   So do we know for a fact that our universe is not closed?

✦   Not really. We can measure the speeds of distant galaxies fairly precisely along our line of sight, based on the amount of their redshift. But unfortunately, we cannot tell very precisely how far away they are. Without being able to relate distance and speed exactly, we cannot tell if they have enough speed to fly apart forever.

✧   But can't we just observe the most distant galaxies and see if in fact they have begun to slow down appreciably?

✦   In principle yes, but in practice the observations are just not accurate enough to tell for sure.

✧   Well, let's just suppose the universe were closed and the big bang had not been quite as big. What would we see?

✦   At some point the big crunch would begin. The universe would reach its maximum size and then start to collapse on itself. The light from distant galaxies would bring us news of the universal collapse through their blueshift.

✧   Big crunch—that sounds like a great name for a breakfast cereal.

What if the big crunch started?

✧   I hope I'd sell all my stocks first. But presumably you mean what if all the matter in the universe reached its maximum expansion and began contracting, with distant galaxies approaching us rather than receding.

✦   Right—which we could tell because distant galaxies would show a blueshift rather than a redshift.

✧   But wouldn't it take billions of years for the light from those distant galaxies to reach us, so that we wouldn't know about the big crunch until long after it started?

✦   No. Actually, we'd know about it much sooner, because if the contraction started at the same time everywhere, we would see blueshifts in nearby galaxies much sooner—maybe only millions of years from now.

✧ I guess that's no time at all in the grand scheme of things. But redshift-smedshift, what difference does it make if the universe is expanding or contracting?

✦ Of course, if the contraction proceeds for the same time as the expansion, the universe will perhaps collapse entirely into a very tiny region. Some cosmologists have, in fact, suggested we are in for an endless series of big bangs each followed by a big crunch.

✧ It sounds like the ultimate recycling program to me.

✦ In fact, getting a bit more speculative, some scientists have suggested that maybe even the laws of nature and the value of physical constants get "recycled" after each of the big bangs.

✧ But what would things be like if the big crunch actually started?

✦ Well, warmer, for one thing. During the expansion of the universe, the cosmic background radiation, a space-filling remnant of the big bang, cooled drastically as the expansion proceeded. So if the expansion became a contraction, this radiation would heat up, eventually making all space aglow with radiation.

✧ The thought makes me all warm and tingly. What else would we observe (as if anyone would be around to do the observing)?

✦ If I may get more speculative again. . . .

✧ Go ahead.

✦ Some scientists have suggested that the cosmic "arrow of time" is tied in with other arrows of time.

✧ Come again?

✦ The idea is that if the universe started contracting instead of expanding, it would be as if the "movie" of the universe had started to run backward.

✧ So in other words, all the processes of nature would start going backward according to this theory.

✦ Right. Dead, cold stars would begin sucking up light from hot space, and expanding spheres of radiated light would converge back onto their sources.

✧ Presumably, the dead would rise up from the ground, people would then live their lives backward, and ultimately be planted in their mothers' wombs for "burial."

✦ Unfortunately, you might not notice any difference from current "forward-time" happenings if your brain processes were also running backward.

What if space weren't completely empty?

✧ Of course, space cannot be completely empty, because stars, planets, and galaxies exist in space, but I suppose you're thinking of the space between all that stuff.

✦ Exactly. A little over a hundred years ago scientists believed that there was a kind of substance they called the "aether" that filled all of space.

✧ Why did they come up with such a crazy idea?

✦ Because they knew that light is a wave, and every other type of wave has to have some kind of material medium in which to travel: water waves travel in water, sound waves travel in air, and so on. They thought that if you had a light wave, there needed to be something that was doing the "waving" when light traveled from the stars through what appeared to be empty space.

✧ When you put it that way, the idea doesn't sound quite so crazy. But if this aether filled all of space, it would have all the properties of empty space, wouldn't it? It would be completely transparent, weigh nothing, and offer no resistance to the passage of planets. How could we ever hope to detect its existence?

✦ By the fact that light traveling through the aether would have a specific speed with respect to it. If our planet happened to be moving through the aether, it would be just as if an aether wind were blowing past us. That aether wind would affect the speed of light in the same way that a regular wind affects the speed of sound.

✧ So what kind of experiment could you do to measure Earth's speed through this aether?

✦ To understand the experiment, think of yourself in an open field with a wind blowing. Let's say you are exactly 20 feet away from reflecting wall, and you use one of those digital tape measures based on ultrasound echoes to measure the wall's distance, based on the round-trip travel time of the ultrasound pulse. The measurement would give the correct result only in the absence of any appreciable wind. If you made a measurement when the wind was gusting in different directions, the result would depend dramatically on the wind direction and speed.

✧ What would I find from these measurements?

✦ For any given wind speed, you would find the longest round-trip time (and hence the largest value recorded as a distance on the tape measure) when the wind blew directly toward or away from the wall.

✧ I guess I can understand that, because if we take a ridiculous extreme with the wind blowing at half the speed of sound, then downwind time is a third less and the upwind time is 50 percent more, so the total time is 16 percent more than the no-wind case. But what did they find when they used this technique in the real experiment to measure the speed of Earth through the supposed aether that fills all space?

✦ Of course, in the real experiment they used light rather than sound, and the "wind" was supposed to be due to Earth's motion through the aether. To everyone's surprise, the experiment showed that the speed of light was the same no matter which way it traveled on Earth—a result which seemed to imply that Earth was at rest in space. Of course, we now know that the real explanation is that there is no such thing as an aether.

✧ But suppose the experiment had, in fact, measured our speed relative to a space-filling aether—what then?

✦ In that case there would be an absolute reference frame from which to judge motion, and the key assumption of relativity theory would be incorrect. Incidentally, even though the aether doesn't exist, there is an "echo" of the aether idea in the cosmic background radiation left over from the big bang.

✧ You'd better explain that one.

✦ Because of Earth's motion through space, the cosmic background radiation coming from one direction has a slightly higher frequency than the cosmic background radiation coming from the opposite direction—the former being blueshifted and the latter redshifted. In effect, we use the observed differences to measure our absolute speed through space.

✧ But isn't this in direct conflict with the theory of relativity, since we can define a "preferred" (absolute) reference frame at rest in space?

✦ No, because the preferred reference frame is taking part in the universal expansion. Effectively, you get a different preferred reference frame at each point in the universe. Think of the molecules of gas in an explosion. A miniature observer sitting on one of those molecules would be in a preferred reference frame only at that point in the gas.

What if Earth were flat?

✧ In this day and age, how can we even imagine such a possibility? We have pictures of the round Earth taken from space by astronauts. Surely you are not suggesting that NASA missions are part of some huge conspiracy to delude the public, and that all the photographs and even the moon landings were concocted in a film studio.

✦ No, but there is a small band of flat-earthers who believe precisely that. I don't know how a flat-earther would try to explain how you can see a shadow of the *round* Earth on the moon during a lunar eclipse. But in any case, I have something else in mind—the idea of an earth that is flat in the same sense as the six sides of a cube are flat.

✧ How could Earth or another planet be in the shape of a cube? Aren't all planets and stars spherical?

✦ It all depends on how large the body is. Have you ever seen photographs of asteroids or comets taken by spacecraft? These small bodies look anything but spherical, and often have the appearance of huge irregular rocks.

✧ But this wouldn't apply to a planet the size of Earth, would it, where the inward force of gravity shapes the body into a sphere?

✦ As you probably know, Earth is not quite spherical. Due to its rotation, it bulges slightly at the equator. Also, mountains and valleys are bumps and depressions on the surface akin to the irregularities of an asteroid—though proportionally very much smaller.

✧ So what would happen if somehow an asteroid or a planet were carved into the shape of a cube by our removing excess matter?

✦ In the case of an asteroid, probably nothing would happen, since gravity on an asteroid is not strong enough to force it into a spherical shape. But in the case of a planet the size of Earth, the carved cube would be short-lived, since gravity would soon restore the planet to a nearly spherical shape.

✧ Exactly how would that work?

✦ Think of yourself standing on the cubical planet. If you stood at the center of one face, "down" would be toward the center of the cube, and the surface would look completely flat and level all the way to an edge. Now picture yourself standing on a corner of the cube. "Down" would now again be toward the center of the cube.

✧ So in effect I'd be standing at the pinnacle of a gigantic pyramid-shaped mountain.

✦ Exactly. In fact, from simple geometry, the height of the "mountain" works out to roughly 73 percent of the radius of the planet. Such a mountain would collapse under its own weight, restoring the spherical shape of the planet.

✧ Well, if a cubic planet the size of Earth could not exist and a cubic asteroid could, I guess there must be some largest possible asteroid or planet where gravity would not be quite strong enough to crush the corners of the cube.

✦ Right. We can deduce what that size would be by using a simple scaling law. On Earth, mountains can reach heights of no more than about $5\frac{1}{2}$ miles before gravity brings them down. Thus, earthly mountains reach a maximum height of 0.14 percent of the planet's radius.

✧ So the corner "mountains" on a cubic planet are in a sense about 521 times larger than mountains on Earth when both heights are expressed as a percent of the radius. But how does the height of mountains depend on the size of a planet?

✦   On a planet half the size of Earth, where gravity would be half as strong, mountains could grow four (2 × 2) times as tall expressed as a fraction of the radius. So by extension, it would require a planet roughly 23 times smaller than Earth for mountains to grow 521 (23 × 23) times taller as a fraction of the radius—just tall enough to correspond to the "mountains" at the corners of a cubic planet.

✧   I think it would be fun to ski down the flat slopes of a cubic planet.

What if you were at the edge of the universe?

✧   What would I see if I stuck my head out beyond the edge? Another universe? Where is the edge, anyway?

✦   Picture yourself sitting on our Milky Way Galaxy—a mere dot in the grand scheme of things—while all the galaxy-dots in the universe fly away from each other as the universe expands. In any direction you look, galaxies are receding faster the farther out you looked. At some distance their speed equals the speed of light—that's the "edge" of the observable universe, about 15 billion light-years away.

✧   But if the edge is equally far in all directions, doesn't that put us at the center? Hasn't the notion that we occupy a special, central place in the universe been abandoned long ago?

✦   Cosmologists today believe that if you were sitting on any galaxy-dot you would see essentially the same thing—so in a sense, observers on each dot could consider themselves at the center.

✧   I could see this for an infinite universe, but how can it be true if the universe is finite?

✦   Let's pretend space has only two dimensions rather than three. Suppose the galaxy-dots are distributed at random on the two-dimensional surface of a spherical balloon whose inflation causes the space between all galaxies to grow.

✧   I sure hope there are no sharp objects around—one big bang is enough. Where is the edge of the universe in this model?

✦   As you sat on your dot-galaxy, you could draw a series of concentric circles on the surface of the inflating balloon. One of these circles would have a radius that would be growing at the speed of light—that would be the edge of your observable universe, because no information could reach you from beyond this limit.

◇   And if I somehow managed to travel to a galaxy located on this circle, I assume that I would see the same thing as I do now, so the edge would still be the same distance away.

✦   Exactly. Traveling to the edge of the universe is a lot like trying to reach the end of the rainbow.

◇   Is all this speculation, or is it backed up by observations of some kind?

✦   We obviously have no information on how the universe might look from a distant galaxy, but we can observe how the universe looks in different directions.

At one time it was thought that the universe should show no large-scale structure, or that it should look more or less the same in all directions (be isotropic). Recent observations show this to be not quite true. If we were wrong about the universe being isotropic, it's possible our belief that the universe looks the same everywhere is also wrong.

✧ So everything you just told me about the universe not having any real edge could be wrong?

✦ Some cosmologists find this possibility very ugly on philosophical grounds, but the question is not one that can be settled by measurements—until we hear from some extraterrestrial astronomers on a galaxy near what we consider to be the edge of the universe, which is not too likely.

What if the universe slowly rotated?

✧ If the universe rotated fast enough, I suppose I'd get dizzy. Actually, it's hard for me even to imagine the rotation of the entire universe. Don't we need to have some outside observer to judge that a rotation is taking place?

✦ Before we worry about that, let's start by considering the rotation of Earth on its axis. One "obvious" piece of evidence for Earth's rotation is the apparent nightly rotation of the heavens about the North Star. Of course, ancient peoples thought it was the heavens that rotated about a stationary Earth.

✧ How can we be so sure they were wrong?

✦ Using the Foucalt pendulum experiment. Picture yourself standing at the north pole holding a swinging pendulum. As the pendulum swings back and forth in a fixed plane, you and Earth rotate; so from your point of view the plane of swings appears to rotate. You can easily get the idea without going to the north pole, just by rotating in a circle while holding a swinging pendulum. You can find Foucalt pendulums in many museums. They offer direct proof of Earth's rotation on its axis.

✧ OK, I guess I have no doubt that Earth is rotating, but what about a somewhat larger scale? How could we tell that our galaxy is rotating, for example—particularly without observing any outside objects?

✦ The galaxy is not a solid object, of course; it is made up of a hundred billion stars, each of which follows its own orbit about the galactic center. The fact that all these orbits are mostly in the same direction makes it somewhat meaningful to talk about the rotation of the galaxy. However, different parts of the galaxy complete a revolution in different times. We can show that the galaxy is rotating by looking at the motions of stars in our region of the galaxy, after we subtract out a random component to their motion.

✧ OK, are we now ready to tackle the rotation of the entire universe?

✦ The universe is different from our galaxy because, unlike the galaxy, it has no massive core that serves as a natural center about which a rotation could occur. On the other hand, if the universe were actually rotating, there would have to be an axis of rotation. In this case, our view of the cosmos would depend on our distance from this axis. If we were located right on the axis, we would also see some kind of systematic variation as we looked in different directions.

✧ What sort of variation?

✦ Think of the rotating Earth and its equatorial bulge, or the flattened disk of the galaxy—both being shape features that arise from the rotation.

✧ I'm not sure what a "bulging equator" of the universe might look like.

✦ Basically, the distribution of galaxies in space, judged from the temperature of the cosmic background radiation, would have to show some kind of systematic variation in direction. Nothing along these lines has so far been seen, but the idea of a rotating universe is not something that can be dismissed out of hand. In fact, why shouldn't the original "cosmic egg" that gave birth to our universe have been rotating at the time of the big bang?

✧ I Einstein know God doesn't play with dice, but perhaps he plays with tops.

What if you fell into a black hole?

✧ I was walking down the street once and almost fell into one. No, I guess that was a manhole. Let's say my spaceship was in orbit about a black hole having a few times the mass of our Sun. Wouldn't the ship gradually get sucked in, no matter what?

✦ Not necessarily. If there were a swirl of matter falling into the hole, the matter might carry you into it like a whirlpool, but otherwise you could safely orbit the black hole indefinitely. Also, you wouldn't need to be continually firing your rocket engines to stay in orbit, any more than Earth needs rocket engines to remain in its orbit around the Sun.

✧ OK, let's say I did fire my rocket engines to leave orbit, but by a tragic mistake I slowed my ship's speed rather than increased it. As a result, the ship no longer had enough speed to remain in orbit, and it began to fall toward the black hole from a distance equal to our current distance from the Sun. What would I observe?

✦ As you fell toward the black hole, the pull of gravity would initially be very small, so your speed would increase slowly. But the closer you got, the stronger the pull would be, and the greater your acceleration toward the hole. It might take as long as 65 days to reach the event horizon of the black hole—the point of no return.

✧ Why is there a point of no return for black holes?

✦ At any distance from the black hole, we can ask what the escape velocity would be at that distance. Obviously, the escape velocity would increase the closer we got to the center of the black hole, reaching the speed of light at the location of the event horizon. Since nothing can travel faster than light, nothing can escape from within the event horizon—which is how it got its name.

✧ What else would I observe as my ship fell toward the event horizon?

✦ Initially, not much. Even though the black hole's gravity would be pulling you inward, you would not feel the force of gravity, as you would be in a state of free fall, or weightlessness. But the closer you got, the more important tidal forces would become. If you were falling in feet first, the gravitational pull on your feet would be greater than that on your head, and you would be stretched out like a piece of taffy. In the unpleasant process, friction would heat your body to millions of degrees, and your body would vaporize.

✧ I'm beginning to regret this trip. Let's say *you* fell into a black hole, and *I* were just watching from a safe distance away. What would I see?

✦ Surprisingly, the scene would look quite different to a remote observer. The theory of relativity predicts that gravity has an effect on time. So while I would find that it took 65 days to reach the event horizon, you, from a safe distance away, would find that time ran more slowly the closer I got to the event horizon. In fact, from your viewpoint I never would quite reach the event horizon but would be inching toward it with continually decreasing speed forever.

✧ I remember reading about black holes being possible gateways to another universe, or possibly to distant points in space or time in our universe. Could black holes be used as time machines if you fell into one?

✦ It has been suggested that rotating black holes might indeed be "worm-holes" to another universe. Even if this theory is correct, you probably would not want to become a piece of taffy, due to the enormous tidal forces, while falling into one. Surprisingly, the more massive a black hole, the smaller its tidal forces, so conceivably, if there are any galaxy-size or bigger black holes, you might survive the trip into one—although you might get a lethal dose of radiation along the way. If you make it, send me a postcard from wherever and whenever you wind up.

Chapter 15

Time

What if time flowed slower in some places than others?

✧   Surely this one is hypothetical—right?

✦   Actually, no. The rate at which clocks keep time do vary with the strength of gravity, according to Einstein's theory of general relativity. So if you climbed to the top floor of a building, where gravity is slightly weaker than on the ground floor, clocks would run slightly faster.

✧   I'm going to move my office to the top floor in that case. Are the differences in time really noticeable for a typical-size building?

✦   Such differences in times have actually been measured—largely because time (or frequency) can be more accurately measured than any other quantity. For example, for every 10 meters increase in elevation above the surface of Earth, gravity weakens by 0.0003 percent, and a clock would run faster by one second in 100 million years.

✧   On second thought, maybe it's not worth moving the office. Anyway, just because clocks are affected by gravity, does it make sense to say that time itself is affected by gravity? After all, clocks can be affected by all sorts of things—pressure, temperature, and so on—which have no effect on time.

✦   When we speak of "clocks," we really are referring to any process whose duration is used to mark off time intervals. According to relativity, the time for all such processes is affected in the same way. Nowadays, most scientists (unlike some philosophers) don't believe there is any "background" time independent of events in the universe. If all clocks are affected by gravity, it seems reasonable to say that time itself has been affected.

✧   Still, the size of the effect you mentioned seems too small to be of any consequence, and I don't intend to get anywhere near a black hole, where stronger gravity would make the effect much more noticeable. Is there some other way that the passage of time could depend on my location?

✦   If we're just talking about your psychological perception of time, that can be affected by all sorts of things, including the time of day, your mood, your temperature, external stimuli (or their lack), and your possible ingestion of mind-altering substances. We might imagine that at certain points on Earth there was something in the air that induced an altered state of consciousness in which time was perceived to pass at a slower or faster rate than normal.

✧ It might be quite interesting to take a "trip" to those places. If everyone's mental processes were speeded up twice as fast as normal, for example, what would they perceive?

✦ Events in the outside world would appear like a movie slowed down to half speed. It would be slowed down, not speeded up, because if your "brain clock" is running twice as fast, any outside event that previously lasted one "tick" of your brain clock now lasts two ticks—so events would seem slowed down.

✧ I'd want to spend my weekends at those places on Earth where my brain clock ran twice as fast as usual. That way the outside world would run twice as slow for me, and my weekends would last twice as long.

✦ That's not so clear. You might much prefer spending your weekends at places where your brain clock ran half its normal rate and the outside world went by at twice its normal speed. That way you could experience twice as much during the weekend, and it would seem twice as long based on all the things you did.

✧ In fact, why not go to a ridiculous extreme and imagine a place where your brain clock ran at one-thousandth its normal rate. If you went there, it would seem like ten years went by each weekend. Boy, would I like to open a resort at a place like that!

✦ There's a slight problem with your scenario. With your brain running at a thousandth its normal speed, events in the outside world would seem to be happening a thousand times faster. You could not possibly cope with the flow of information coming in. Just imagine the hopeless task of trying to watch a movie running a thousand times faster. The only thing in the outside world you would be able to notice would be those events that now occur over too long a timescale for us to see anything happening. If you enjoyed watching your hair grow or the sun zip across the sky, this hotel might be for you.

✧ Come to think of it, spending a weekend at a hotel where my brain was speeded up a thousand times would pose similar problems. As far as I was concerned, the outside world would be like a movie running a thousand times slower—nothing would appear to be happening in the outside world.

✦ Quite the contrary. Nothing would be happening on our normal timescale, and your body would appear almost frozen still. But you'd be able to directly witness all kinds of phenomena that now occur in an instant, and see them in extreme slow motion. Observing hummingbirds move their wings, water balloons exploding, and sound waves created by leisurely vibrating objects would be a wonder to behold.

What if we could travel into the future?

✧ I guess you could say we can travel into the future right now. We do it constantly at a rate of one second per second. . . . There, I just did it again. But I don't suppose that's what you mean.

✦ Right, I mean literally traveling into the future, to whatever year you chose.

✧ But wouldn't that require a time machine?

✦ Actually, a rocket ship would do nicely, although it would need to be a rocket much in advance of anything on the drawing boards today, because it would need to be capable of attaining speeds very close to the speed of light.

✧ This, I assume, has to do with Einstein's theory of relativity, which says that moving clocks run slow.

✦ Right. And it's not just clocks that slow down but any measure of time, including biological, psychological, chemical, or physical processes—everything.

✧ I don't see why that should be, but that's probably the subject of another essay. This is just a theory, isn't it? Has it actually been observed?

✦ Sure. Using very accurate clocks aboard a high-speed jet aircraft, it has been found that they run slow compared with other clocks on the ground by exactly the amount the theory of relativity predicts.

✧ Why can the effect only be seen with very accurate clocks?

✦ Because even the speed of a jet aircraft is very small compared with the speed of light. Unless the speed is close to the speed of light, the size of the effect is very small, so an accurate clock is needed.

✧ OK, so how do we use this effect to travel into the future?

✦ Let's say you got into a rocket ship that over a long period of time accelerated very close to the speed of light and traveled to a star 50 light-years away.

✧ Presumably, that means that light from the star would take 50 years to reach us, so I should reach the star in just over 50 years if I moved close to the speed of light—right?

✦ That's 50 years Earth time. For you in the ship time would run more slowly than on Earth—how much more slowly would depend on just how fast you were moving. It could well be that if you traveled fast enough, it would take only a year for you to reach the star and return—compared with the 100 years the round-trip would take according to Earth clocks.

✧ So on my return from my one-year journey, I would really be visiting Earth 100 years in the future, or any other time, depending on my speed.

✦ Right, but forget about going back in time by taking a round-trip. We'll talk about that another time.

✧ OK, but there's one thing I don't get. According to relativity theory, why couldn't I in my rocket ship consider myself to be at rest. In that case, it's Earth that would be moving, and so Earth clocks should be the ones that run slow, not mine.

✦ You just asked about a problem known as the twin paradox, so named because we can imagine one twin staying behind on Earth and another going on a round-trip spaceship journey.

✧ OK, so why can't each twin simply say the other one's time runs slow? In what way is the situation not symmetrical between the two twins?

On returning to Earth, the space traveler is surprised to find that he missed not only his son's graduation but most of his life as well.

✦ It's not symmetrical because the spaceship twin would not consider himself to be at rest. Even without looking out the window, he could sense the acceleration and deceleration of the ship, and know that he was the one who was moving.

What if motion affected clocks in different ways?

✧ Just because one type of clock runs slow due to motion, I don't see why every other type of clock (including my brain clock) should also run slow by exactly the same amount.

✦ That's exactly what we did assume in the previous essay. Relativity theory says that all clocks must run slow in exactly the same way, and we now want to imagine what the world would be like if this were not so.

✧ OK, let's imagine that I gather a whole bunch of different types of clocks in my spaceship, including some "biological clocks," such as us. We can measure biological time by such means as the steadily growing length of our fingernails and hair, and our direct perception of time.

✦ You need to have some way of telling whether or not all the clocks in your ship are keeping the same time. You could just listen to the chimes when all the clocks sounded the hour, but that might drive you cuckoo, so instead let's imagine all the clocks hooked into a computer (with its own clock). The computer also monitors all the biological clocks (including the length of your fingernails) and continually prints out a message that the clocks are "in synch" or "not in synch."

✧ Presumably, when my spaceship was at rest, all the clocks would remain in synch, assuming I had assembled a set of good clocks. How would they look to you if you were zooming past my ship in your own spaceship?

✦ I wouldn't even have to try it to tell. The various clocks would have to remain in synch, because I would have to agree with you as to whether the computer printed out the "in synch" or "not in synch" message, which is a permanent record of what happened.

✧ So does this imaginary experiment prove it is impossible for motion to affect clocks differently depending on the type of clock?

✦ No, it only shows that the motion of the observer cannot make any difference. We have not yet considered the effect of any motion of your spaceship through space.

✧ But I thought that according to relativity it doesn't matter which one is moving, and that only relative motion counts?

✦ Right, but we're now considering a situation that conflicts with relativity (clocks affected by motion differently), so we don't want to assume anything that is part of relativity.

✧ OK, let's suppose that when my ship was parked on Earth the clocks were always in synch, but when I traveled at high speed through space they were no longer in synch. What could I conclude based on that observation?

♦ You could conclude that Earth was in a "special" frame of reference, which you could define as being at rest in space. You could also conclude that the idea of uniform motion only being relative—the central idea of relativity—was wrong, because there would be a clear difference between someone on Earth watching your ship go past and someone going past Earth watching your parked ship. Only in the latter case would the clocks in your ship all be in synch.

What if you stepped into the fourth dimension?

✧ As everybody knows, the fourth dimension is time, or at least that's what relativity says, but I've never understood why that should be.

♦ OK, let's consider an object, such as yourself. You have the three spatial dimensions of height, width, and breadth, but you also have the dimension of duration—how long you last.

✧ Sure, but couldn't I just as well take temperature to be my fourth dimension? What's so special about time?

♦ Actually, there is something special about time in the way it combines with the three dimensions of space. But to see how this works, let's start with a different question. Suppose you never moved about in three dimensions, and the only thing you ever experienced of the world was what you viewed on a movie screen. Could you somehow figure out that the pictures seen on the flat screen were actually of three-dimensional objects?

✧ The screen images are obviously two-dimensional, so how could we ever learn about the third dimension by watching them?

♦ When a three-dimensional object rotates and it is viewed from different directions, its appearance changes in a way that tells us about its three-dimensional shape. Imagine a cube, for example, rotating in front of you with its image projected on a screen. In the screen image of the cube, one edge appears to grow and another shrink in a very predictable way as it rotates. Based on these changes, we could figure out that we were looking at a rotating three-dimensional object, rather than something changing its shape.

✧ So what has all this got to do with time as the fourth dimension, and how could an object "rotate" in four dimensions?

♦ According to relativity, the equations describing a rotation in four dimensions are almost identical with the changes in the space and time dimensions of a moving object. In fact, the four dimensionality of space-time is one way to explain relativity.

✧   Time seems so obviously different from space, it seems strange to tack it on as simply one additional dimension. How about the possibility of a fourth dimension to space itself? Is that simply science fiction?

✦   Not at all. According to modern theories of particle physics, there may be ten or eleven dimensions in the universe—we just aren't aware of them, because they are all "curled up."

✧   I'm not sure what it means to have extra dimensions curled up.

✦   Since we can't visualize the situation, let's take one we can. Suppose we were two-dimensional creatures living on the surface of a long cylinder. We would, of course, have no awareness of the three-dimensional world outside the cylinder. Now suppose that the radius of the cylinder were to shrink to zero. One of the dimensions of our universe would have "curled up" to zero, leaving us in a one-dimensional world we might call Line Land.

✧   In science fiction stories, characters sometimes step into another dimension and return. What might happen if we stepped into a fourth dimension of space and returned?

✦   Picture again what it would be like to live on a two-dimensional world, such as the surface of a plane, cylinder, or sphere. To be specific, let's say there were two types of creatures—the righties and the lefties, who looked just like animated drawings of right and left hands. If a rightie could somehow leave the two-dimensional surface and flip over before coming back, it would find itself transformed to a lefty on its return.

✧   So based on this analogy, what might happen to me if I took a brief trip into the fourth dimension?

✦   You might come back in mirror-reversed form, with your heart and all your other organs on the opposite side of your body, with left and right completely interchanged.

✧   ?ti did I fi llet I dlouc woH

What if you could
travel into
the past?

✧   Now we're getting really weird. This is strictly science fiction, right?

✦   Maybe, maybe not.

✧   If we could travel back to the past, especially our own past, we could change all sorts of things—like maybe I could have taken Lucy Kringle up on her offer in seventh grade.

✦ Most people might want to visit all sorts of great moments in our history.

✧ Like the Sermon on the Mount, Napoleon at Waterloo, the Kennedy assassination, and Elvis's last concert. But if time travel is discovered in the future, won't there be hordes of time tourists going to the more popular events?

✦ Probably, and in fact the number in attendance would equal the number of future time travelers for all time. The fact that things would get so crowded has suggested to some people that time travel is impossible.

✧ What do you mean?

✦ Even if time travelers did manage to blend in with the crowd, if so many of them had been present at some historical events, surely the size of the crowd would have been noted in the historical records.

✧ Well, maybe the time-travel agency that books these trips has a rule about only going back to events where there was a big crowd. But aren't there are all kinds of other problems and paradoxes with the idea of time travel?

✦ Yes. For one, going back to the past opens up the possibility of changing the past, possibly causing profound changes in the present as a result.

✧ Like my going back and accidentally killing my own grandfather, which would prevent me from existing!

✦ I don't know why you time travelers always pick on your grandfathers. Why not go back and kill your grandmother, one of your parents, or even your earlier self.

✧ I really wouldn't want to do that to old grandpa, but the very fact that I could raises serious doubts about whether time travel to the past is possible.

✦ Maybe not. One way out is to say that you could travel to the past and observe it, but not affect anything—almost as if you were observing a movie. Another way would be to say that you could affect things but not change anything. So if you went back in time and tried to kill your grandfather, something would always prevent you from doing it no matter what you tried. In fact, if you did go back and actually "succeeded" in killing grandpa, it would have to turn out in the end that the man you killed really wasn't your grandpa after all.

✧ I don't see why, if I went back really determined to do it no matter what the cost and made foolproof preparations, I couldn't carry out the dastardly deed.

✦ Because there is only one past, and if you did succeed in killing a specific individual, it would be part of the historical record. If, in fact, it was grandpa you killed, you would have to have done it after he started your mother or father.

✧ So basically, any changes made to the past would have to leave the present unaffected.

✦ Actually, some science fiction writers have claimed just the opposite. In one story, the most insignificant change in the past—a time-traveler accidentally stepping on an insect—was supposed to have caused a ripple of ever-increasing consequences that eventually had a catastrophic effect on the present. The idea is much like the "butterfly effect" associated with the theory of chaos, in which even the flapping of a butterfly's wings on the other side of the planet could, in principle, affect the weather here.

✧ Does time travel have the slightest possibility of being possible?

✦ As described in another essay, it is well established that the rate at which time flows is affected by the presence of strong gravitational fields. The ultimate distortion of space and time occurs in the vicinity of black holes, which usually result from the collapse of massive stars. Some physicists believe that if you fell into a supermassive rotating black hole, or a "wormhole," you might come out at some other point in space and time.

✧ I thought black holes were rather nasty things to fall into.

✦ Yes, normally they are. Approaching a typical black hole, you would likely be torn apart by tidal forces even before you fell in. But for a really massive one, it is conceivable you could survive the trip to somewhere and somewhen.

What if you went back in time to ancient Greece?

✧ If I went back, say, to ancient Greece, I imagine I'd be hailed as a genius. We know so much more today about how the world works than ever before.

✦ Before you pack your bags, you might want to think about what it is we "know" that you would be able to convince the ancient Greeks about, without their thinking you were a lunatic.

✧ Well, the ancients were convinced that the Sun and all the planets went around Earth, but now we know this is not true. But you're right; I haven't a clue as to how I would convince them that Earth goes around the Sun.

✦ Don't feel too bad, because without a good telescope, even an astronomer couldn't make the case. In fact most of our knowledge today relies on observations that require special apparatus. Try to convince the Greeks that germs cause disease or that everything is made of atoms, without a good microscope, and you'll see what I mean.

✧ But didn't the ancient Greek philosopher Democritus invent the atomic theory? He apparently didn't require any fancy gadgets to come up with the idea.

✦ True. But only by making observations can we tell whether our theories correspond to the real world. Democritus may have come up with the idea of atoms, but it was not until the twentieth century that their reality was demonstrated.[1]

✧ I guess I might convince the Greeks of today's superior knowledge by building something they had never seen before, or even dreamed of.

✦ That is a pretty big if. Most of the simple technological advances require some material or some component that the Greeks wouldn't have had. You might be able to make a very primitive battery and lightbulb, for example, but almost certainly not one that would glow visibly for any length of time. The technique for making wire is tricky—especially the kind of fine, high-melting-point tungsten wire used in today's lightbulbs.

✧ Surely there would be something I could make that would impress them.

✦ Your best bet might be to try to grind lenses. With lenses you might, with a little experimentation, figure out how to make primitive telescopes and microscopes. The ancient Greeks would be astounded to see the moons of Jupiter, the ring around Saturn, the craters of the Moon, and planets beyond the five they knew about. They would be equally astounded to see the microworld that your microscope opened up.

✧ Anything else?

✦ If you did some research before you traveled back in time, you might be able to build a primitive pendulum clock driven by wooden gears. Even if you couldn't master the escapement and gear mechanisms of a clock, you could certainly have some poor slaves keep count of the pendulum swings and measure time that way.[2] You might even be able to build a primitive steam engine, and show them how to make soap and gunpowder.

✧ I suspect that the Greeks might be better off if they were visited by a time-traveler from the 1800s than one from today.

✦ That's certainly the case. In previous times, people had a much better understanding of their technology, but as a technology matures, that becomes less and less feasible. In the early part of this century, many teenagers used to build their

---

1. One way you can show that atoms exist, even without seeing them directly, is from the "random walks" of tiny pollen particles suspended in a liquid, as a result of the random collisions of atoms with the particles. This Brownian motion (named after the botanist Robert Brown) can easily be seen under a low-power microscope. But without the necessary theoretical arguments, first put forth by Einstein, you probably wouldn't convince the Greeks that Brownian motion was evidence for the reality of atoms.

2. Ctesibus of Alexandria actually did develop a water-powered mechanical clock around 135 B.C., but it wouldn't have kept time nearly as well as one based on the swings of a pendulum.

own crystal radios; far fewer would have any inclination to do it today. The same could be said about automobiles, before their engines became the computer-controlled complexities of today.

✧   But aside from going back in time to help out the ancient Greeks, why does it really matter if people don't have a clue as to how their technology works?

✦   Partly because the fewer the people who understand a technology, the more dependent we become on that handful of people to keep things on track. Also,

Ideas ahead of their time are seldom accepted.

some people might be reassured if they knew that we could reconstruct our civilization following some kind of cataclysm, such as a nuclear war.

✧   Right, and others would be even happier to know that we couldn't. Why else does it matter if people don't know how technology works?

✦   Mainly because they become less and less capable of distinguishing science from pseudoscience, "magic" from illusions, and the possible from the impossible. In today's world of widespread belief in UFOs, alien abductions, astrology, and extrasensory perception, anything may seem possible. In such an environment, we may more easily be induced by charlatans to embark on fantastic, extremely expensive solutions to unsolvable or nonexistent problems.

What if the universe were made into a TV movie?

✧   By Universal Films, I assume. Let's see, the typical TV movie runs two hours, but taking out the advertisements, we're talking about 100 minutes of actual movie in which we want to include the entire history of the universe, which runs . . . ?

✦   Somewhere between 10 and 20 billion years. This estimate comes from noticing that all the galaxies seem to be flying apart from one another—the result of that cosmic explosion we call the big bang. Somewhere between 10 and 20 billion years ago, the galaxies would have all been together at one point. We can't be more exact than that, because the age depends on our very imprecise estimate of how far the most distant galaxies are.

✧   I'd ask where that one point is that the galaxies all came from, but I'll bet that's the subject of another essay. In any case, to be definite, let's say the universe is 15 billion years old. At 16 frames per second, that means that each frame of our 100-minute movie represents about 156,000 years of real time. Does much happen during that first frame?

✦   You might want to close your eyes when the universal movie begins. The movie would probably merit a rating of X or at least R, given the tremendous violence it contains, and the most extreme violence happens in the very first frame. At very early times, the universe was extremely hot. It wasn't until after the first 4 seconds that the universe was cool enough for the protons, neutrons, and electrons of which the universe is made to exist stably. Before then, it was so hot that they would form as often as they would break apart due to the violent collisions.

✧   When did these particles get together to form atoms?

✦   The vast majority of atoms in the universe are hydrogen. Electrons and protons got together to form hydrogen atoms when the universe was about a million years old—6 frames, or a third of a second, into the movie. At that point the universe had cooled to about 3,000 degrees.

✧ How about the vast collections of stars we call galaxies—when did they form?

✦ About a billion years after the big bang, or about 7 minutes into our 100-minute movie. We know that because looking out in space is essentially looking back in time, and we don't see any galaxies closer than about 7 percent of the distance to the "edge" of the universe.

✧ And Earth and solar system—when do they make their appearance in the movie?

✦ Not until the movie is more than two-thirds over—at about the 70-minute point. At this time the debris left over from some earlier exploded stars got recycled to form our solar system.

✧ What about the *interesting* part of the movie?

✦ You're probably talking about the first sex scenes. The earliest life-forms on Earth appear to have begun around 3.6 billion years ago—76 minutes into the movie (still no sex). The first creatures that crawled from the seas onto land didn't appear until about 425 million years ago—the last 3 minutes of the movie—and the earliest mammals didn't appear until the last minute and a half.

✧ How about people?

✦ Our earliest apelike ancestors came onstage 25 million years ago, during the last 10 seconds of the movie, and the first true humans didn't appear until the last second. All of recorded history (about 5,000 years) is contained in the last two-thousandths of a second.

✧ Talk about being a bit player! I think we need a sequel.

What if time were to flow backward?

✧ I'm not quite clear on how we know which direction time is "flowing." Would backward-flowing time mean that the future occurred before the past, or some such nonsense?

✦ To avoid confusion let's rephrase the question as: what if events occurred in backward time order, like events in a movie run in reverse?

✧ Whenever I've seen movies played backward, there was no doubt that the time-reversed events could never occur in real life: things like eggs unscrambling, shattered objects reassembling, and hair clippings rejoining someone's head.

✦ A strange thing about all these events that occur only in one time-direction in real life is that the basic laws of nature (with one possible subatomic particle exception) do not distinguish one direction of time from the other. For example, suppose a pair of billiard balls make an elastic collision—one in which no mechanical energy is converted to heat. Newton's laws apply to the time-reversed collision just as well as the original one. Watching a movie of just two balls colliding, you could not tell whether the film was running backward or forward.

✧ Sure, but if I saw a movie in which the cue ball collided with 15 colored balls arranged in a triangle, and scattered them, I could surely tell if that movie were run backward.

✦ True. But it's all a matter of probabilities. If you somehow managed to give the 15 scattered balls precisely the right velocities and directions, in principle they could all collide in such a way as to form a static triangle, with the cue ball flying off—it's just extremely unlikely to have an initial condition that would give this result. Other supposed examples of events that go only in one direction are also a matter of likelihood of the initial condition.

✧ Could we consider some more examples? How about an expanding circular water wave created when a pebble drops into the middle of a pond?

✦ To have the time-reversed situation (a contracting circular wave that converges to a point and lifts the pebble) is something you'll certainly never see in nature. But in principle, if you had some way of starting waves all along the shoreline at exactly the right times (depending on each point of the shore's distance from the center of the pond), you could create a converging wave. In fact, you can easily see converging circular waves if you put some water into a glass pie plate and dunk your finger repeatedly in the center. The outward waves become inward waves when they are reflected.

✧ I assume the same could be said for any other situation involving expanding waves, such as the light given off by a star, the seismic waves from an earthquake, or the sound from my voice. In each of these cases, we could imagine the process running in reverse if the peculiar initial conditions were established. In such a universe, stars would suck up light, our vocal cords would suck up sound, and so on. Are there any other processes in nature that appear to single out a direction in time?

✦ Another would be the expansion of the universe. Here again, it's strictly a matter of the initial condition (in this case the big bang). However, at some point in the future, the universe may begin contracting—reversing the time sequence.

✧ There is one distinction between past and future that is completely independent of events in the outside universe, namely, our internal mental sense of time. We very rarely have any confusion between whether events belong to the past or future—leaving aside déjà vu feelings. Our mental perception of time is that of a "now" that slides forward at a uniform rate, ever enlarging the amount of time in the past and decreasing the amount in the future.

✦ Strangely, there is nothing in physics corresponding to this psychological perception. In fact, it seems likely that if events in the external world were to run backward, as well as our own mental processes, we might perceive things much as we do now.

✦ Almost certainly not. Communicating with someone whose time sense is opposite yours means that your answer to their question arrives before you hear the question.

✧ What if there were a "parallel universe" in which time went backward, including the mental processes of its people? Could we communicate with them?

What if an object could be in two places at the same time?

✧ Short of having an out-of-body experience, I can't imagine how that could happen.

✦ It could happen routinely if time travel were possible—which is, of course, quite a big if! To get the idea, imagine that you wanted to take a trip from city A to city B, located 120 miles away. Let's say you started out on your journey at noon and drove at 60 miles per hour. At the end of an hour you'd be at the halfway mark, and then "oooeeee. . . . "

✧ What's that?

✦ Twilight-zone music. Time starts flowing backward at that point, so if you continued moving toward city B, you'd arrive there at noon—the same time you started your journey—putting you in two places at the same time. In fact, for any time between noon and one o'clock, there would be two points on the line between cities A and B where you could be found.

✧ I suppose this example shows that the idea of being in two places at the same time is just as absurd as backward time travel. But could either of these things actually occur?

✦ Physicists believe they can. The trip we just described, in which you went forward in time and then backward, is believed to happen all the time among electrons and other subatomic particles. In fact, the particle we call the positron (the antiparticle of the electron) is believed to be an electron traveling backward in time. It is possible to describe your journey in the language of particle and antiparticle, without ever referring to time flowing backward, in the following way. We would say that you and your antimatter twin ("ouy"?) started out at cities A and B at noon, traveled toward each other, and met at the midpoint at one o'clock.

✧   What happens when I meet up with my antimatter twin? And how come we no longer need to speak of time flowing backward?

✦   When you and ouy meet, the two of you would annihilate each other, producing pure energy. We would no longer need to speak about time flowing backward, because ouy traveling from city B to the midpoint in forward time *would be actually be you going from the midpoint to city B in backward time.* In fact, this second way of describing the journey is the only way a backward time-traveling you or electron could possibly appear to us, since our minds can perceive time evolving only in one direction, which we call forward.

✧   I'm not sure I care for all this time travel stuff. Isn't there some other way I could be in two places at the same time?

✦   Yes. We normally find it hard to conceive of being in two places at once, because we think of objects such as electrons or people as occupying well-defined positions in space. But in fact, because all matter has a wave nature and waves can be spread out in space, there are ways to observe particles being in two places at once—the two slit-experiment being the most famous example.

✧   What's the basic idea?

✦   Let's say a bunch of water waves are heading toward a long barrier that has two small closely spaced openings (slits). When the waves pass through such a narrow opening, they bend and emerge as expanding semicircles, as if emitted from one point. The waves passing through each opening overlap the waves emerging from the other opening. This overlap of the two sets of waves produces an interference pattern, which means that along certain directions the two sets of waves reinforce one another, and along other directions they cancel each other out. But if you cover one of the openings the pattern disappears, because its existence requires waves passing through both openings at once.

✧   So what has all this to do with an electron being in two places at once?

✦   The two-slit experiment can also be done using a beam of electrons, al-though it is a bit trickier for technical reasons. In principle, you would find exactly the same result as with water waves, even if you were to use a weak beam with very few electrons in it. Since the interference pattern can occur only when waves pass through both slits at once, the only way to get such a pattern with a weak electron beam would be if the individual electrons in the beam, in effect, passed through both slits at once.

✧   But couldn't I put some kind of instrument next to each slit to record unob-trusively which slit an electron actually went through?

✦   If you tried that, you would find that any such measurement would unavoid-ably destroy the interference pattern. In effect, by making the measurement *you* would force the electron to go through only one slit, but if you made no mea-surement it would go through both—as evidenced by the interference pattern.

✧ How come this spooky stuff about being in two places at the same time only happens with electrons and not with people? If it did happen with people, of course, defense attorneys would have a much tougher time trying to establish alibis.

✦ The wavelength of an object can be shown to be inversely proportional to its mass. This means that your wavelength compared with that of an electron moving at the same speed is 10^{35} times smaller—which is the ratio of your mass to that of an electron. Electrons normally have a pretty small wavelength. So if yours is smaller by such a huge factor, there would be essentially no prospect of observing your wave nature in an interference experiment.

✧ That's good. Somehow, the prospect of canceling myself out due to interference appeals to me no more than the idea of annihilating my antimatter twin.

Chapter 16

Miscellaneous absurdities

What if black were white?

✧   This one is pretty silly. Surely you don't mean what if white objects were called black, and vice versa?

✦   No. That wouldn't be particularly interesting.

✧   It also wouldn't be particularly meaningful to ask what if people perceived white as black, and vice versa, would it? For example, I am partially color-blind and am constantly being asked by people, what does red actually look like to me? They are always at a loss when I ask them to first tell me what red looks like to them. How can such questions possibly be answered without letting people inside one another's heads?

✦   You're right. It wouldn't be meaningful to ask what if people always perceived black as white, and white as black.

✧   I think maybe you gave away the game in saying "always." Presumably what you're after is cases where sometimes black is perceived as white, and white as black, right?

✦   Now you're onto something interesting. Let's think about the three things the color of an object depends on: its surface composition, the eye or mind of the perceiver, and the. . . .

✧   Of course, the light illuminating the object. I suppose white objects would appear black if viewed in a dark room, and black objects might appear white when viewed under extremely bright lights, but that also seems something of a cheat.

✦   Right. What we're looking for is a situation where white objects appear black and black objects appear white when the illumination is held fixed for both.

✧   I'm afraid that I'm stumped. Give me a clue.

✦   Have you ever noticed that on a bright sunlit day, if you look at a dark-colored house, the light interior seen through a window appears completely black, and much darker than the exterior?

✧ I'm not sure, but let's try something similar. Here is a small cubical cardboard box that can represent the house. I'll cut a half-inch diameter hole in the middle of one face. Let's view the box in ordinary room light without shining a light directly into the hole. Wow, sure enough, the hole appears completely black, even though the interior of the box is in fact white.

✦ Yes, and if you covered the exterior of the box with black paper, it would appear much lighter ("white") in comparison to the hole when the box was illuminated by a very bright light.

✧ Why should the hole appear so dark, when it is being illuminated by the same light shining on the exterior?

✦ Because how dark something appears depends not only on the brightness of the light that illuminates it but also on the amount of light reaching your eye. Most of the light entering the hole bounces around the inside of the box, where it eventually gets absorbed by the white walls, so very little gets out.

✧ Right, and the little light that does get out of the hole is headed in all directions, so very little of it reaches my eye, and the hole appears black. I guess the reason I made a half-inch hole is that if I made it a lot bigger, too much light would get out, so the hole would no longer appear black.

✦ You might want to see just how big the hole can be made and still appear black, and also try the same experiment with a box twice as big. I think you'll find that the maximum size is proportional to the size of the box, since the fraction of light getting out the hole depends on the ratio of the hole and box sizes.

✧ So I guess you could say that the subject of black and white is not always black-and-white.

✦ Ain't that the hole truth.

What if a picture were worth a thousand words?

✧ I've seen some modern art in museums that was worth much less—all-black paintings, for example.

✦ That's exactly right. Some pictures can be described in very few words (like "all black"), if they have very little information content. A complex picture, of course, can take much more than 1,000 words to describe.

✧ How complex a picture could we describe if we were limited to 1,000 words?

✦ To answer that, we would need to quantify the concept of information content, as it applies to words and pictures. To get the idea, let's consider the picture of the author shown in figure 1. This particular picture consists of an array of $32 \times 32 = 1{,}024$ little squares (pixels), any one of which has a constant shading. If each pixel had ten possible darkness levels, we could specify this or any similar picture made up from 32×32 pixels as a 1,024-digit sequence.

✧ I see. So for example, the number 234517800 . . . would tell us the darkness of each pixel as we scanned across the picture starting from the upper left corner of the image. But how many "words" long would this picture be?

✦ Let's say there are very roughly 10,000 English words, so that any particular word could be identified by a number from 0000 to 9999. The 1,024-digit sequence could then be converted to 256 four-digit sequences, or 256 "words." A picture of the author with four times as much detail as figure 1 would be "worth" around 1,000 words.

✧ I suppose the easiest way to get four times as much detail would be to use twice as many pixels in each direction, giving a 64×64 pixel array.

✦ That's right. Notice, by the way, how the coarse pixels of figure 1 produce a much better image when it is viewed from a distance, so that the pixels appear smaller on the retina. The same effect can be achieved by shrinking all pixels but otherwise making no change in the figure—see figure 2. The reason that figure 2 is more easily recognized as a face is that its image on the retina better approximates the continuous images of the kind we are used to seeing in the natural world.

✧ But there's something I don't understand. Didn't we begin this discussion by observing that the information content of a picture depends on its complexity? The procedure you just suggested for creating a 1,000-word picture would be the same even if every pixel were black—which clearly represents a case of very little information content. As we noted previously, it would only take the two words "all black" to completely specify such a picture.

✦ Good point. Once an image is represented using pixels, we can evaluate its information content by seeing how much the image can be "compressed." We can then define the information content of the picture in terms of the number of digits or words in the compressed picture.

✧ Maybe this would all become clear if we took a specific example.

✦ OK, let's say we had a 32×32 pixel image that consisted only of a single black horizontal line five pixels long at the upper left corner of the picture. If we used 10 shades of gray, ranging from white = 0 to black = 9, we could represent the image by a sequence of 1,024 digits that began 99999000000000 . . . , with zeros filling out the remaining digits. Clearly, it would be very inefficient to represent the image this way.

✧ How could we compress this image to find its true information content?

✦ We need only give the length of each string of like digits that make up the entire number. Since the picture we were considering was a very simple one, it might be coded as 5(9)1019(0), meaning 5 nines followed by 1,019 zeros, which requires only 7 digits to specify, rather than the original 1,024.

Fig. 1. Photo of the author using 32 × 32 pixels. Note how the picture becomes "clearer" when viewed from a distance.

Fig. 2. Photo of the author using 32 × 32 pixels. Note how much clearer the reduced picture is than the larger version in figure 1, even though both are exactly the same figure.

✧   Using your compression scheme, how would we represent a picture consisting of a checkerboard of alternating black and white squares, which could be coded as 90909090909090909 . . . ?

✦   You happened to pick an image for which my suggested compression scheme would not be appropriate. A much more efficient scheme for compressing a checkerboard image would take advantage of the repetitive character of the image. For example, suppose we use the notation $RN()$ to stand for "repeat the string in parenthesis N times." The checkerboard image could then be represented by R512(90). Obviously, the efficiency of a particular coding scheme to represent an image can depend on the nature of the image.

✧   I suppose it's not surprising that a picture that looks as simple as a checkerboard contains very little information content, and can be represented by a simple code.

✦   True, but the human eye is actually not a very good judge of the information content of a picture. Some pictures—especially those of shapes known as fractals—appear exceedingly complex visually, yet their information content, based on the minimum length of a computer code needed to generate them, can be surprisingly short.

✧   I see, or maybe I should say R745 G85 Q496745375!

What if you had a library of all possible books?

✧   What do you mean by all "possible" books? How could you have a library that included books that had not yet been written?

✦   In a short story, "The Universal Library," by Kurd Lasswitz, the author proposed a way to do exactly that. A typical page of a book has roughly 2,000 characters. If we assume a length of 500 pages, that gives us. . . .

✧   A million characters per book. I suppose any book that actually had more than this number of characters could be divided into several volumes. But how many possible characters are there?

✦   We have 26 letters, plus the numbers, plus all sorts of special symbols. For the sake of round numbers let's say we could get by with 100 different characters, including the blank space as a character.

✧   So based on these assumptions, how many possible different books could be written if each of a million characters could take on 100 different values?

✦   It would be 100 times itself a million times, or 100 to the millionth power. This is also equivalent to 10 raised to the two-millionth power (a one with two million zeros after it). Of course, the vast preponderance of these "books" would contain complete and utter nonsense. A typical book might begin: 2!5che

4#&^@-ƒC@@. . . . Once in a while, you might find strings of three or four letters that spelled out English words, and once in a very long while, you might even find a sentence.

✧ But surely there are a very large number of meaningful books that have been written, or could be written.

✦ Yes, but when they are expressed as a fraction of 10 raised to the two-millionth power, they comprise a vanishingly tiny number of cases. For example, there would be only one copy of *Moby Dick* in the universal library, but you would find an extremely large number of books that started out as *Moby Dick* on the first page and then became complete gibberish after that. In fact for any one real book, like *Moby Dick*, there would be a huge number of *Moby Dick* look-alikes, with varying numbers of typographical errors.

✧ I guess, in that case, it doesn't sound as if this library would be terribly useful.

✦ That's exactly right. You might start out reading a sensible account of the history of Western civilization in some particular volume, and then suddenly find you were reading a cookbook or, more likely, complete nonsense.

✧ How much space would this library take up?

✦ Ten to the two-millionth power, though perfectly finite, is such a large number that it is difficult even to think about. If every book in the library were reduced to the size of a proton (about a ten-trillionth of a centimeter), we would still have room only for an insignificant fraction of the library. The number of protons that could fit into the entire known universe is 10^{120}. The number of books left out in this case would be 10 raised to the 1,999,880th power instead of the 2,000,000th power. We would have left out virtually the entire library!

Chapter 17

Make up
your own
what ifs

Thinking up your own *what ifs*? can be a lot of fun—particularly for interesting conversations on long car trips. You might even make a game of it. The game would be to suggest what might follow if the *what if* were true, and perhaps under what circumstances it might be true. To give you a head start, here are a number of topics I considered for possible essays but for one reason or another didn't include.

What if . . .

Aliens 10,000 years ahead of us were contacted?

All days had just one hour of daylight?

Animal laboratory research was banned?

Animals, like trees, had no front or back?

Animals had three legs?

Another religion's God was the true God?

Anyone could build an atomic bomb?

Anything was legal one day a year?

Artistic expression was forbidden?

Astrology was true?

Atoms were really inhabited miniature solar systems?

Bacteria didn't exist?

Beauty and intelligence were inversely related?

Books were no longer read?

Child rearing required a license?

Children could divorce parents?

Clocks had not been invented?

Computers keep getting smarter?

Convicted criminals had to wear signs?

Crime was considered a disease?

Discrimination on the basis of intelligence was banned?

Dreams and reality could not be distinguished?

Dreams predicted the future?

Drugs became legal?

Earth had a Saturn-like ring?

Earth had no magnetic field?

Earth was actually only 5,000 years old, as the Bible says?

Earth was back in its infancy (a few million years old)?

Earth's orbit was highly eccentric?

Eating food was considered a disgusting activity?

Electricity had not been discovered?

Electrons were not all identical?

Energy was nearly free?

Everyone had an identical twin?

Everyone had a photographic memory?

Everyone had terrible body odor?

Everyone was armed?

Everything doubled in size overnight?

Everything not forbidden was compulsory?

False memories could be implanted?

Families were allowed to have no more than two kids?

Faster-than-light speeds were possible?

"Feelies" were developed?

Female babies outnumbered males ten to one?

"Flubber" existed—a material that bounces higher each bounce?

Ghosts existed?

God had a great sense of humor?

Gold or diamonds were abundant on Earth?

Gravity repelled rather than attracted?

Homes were all bugged by Big Brother?

Homosexuality was shown to be genetic?

Human females went into heat?

Human hands had more than five fingers?

Human speech was through birdcalls?

Human vision was only black and white?

Insects could talk?

Insects grew to human size?

Irresistible forces encountered immovable objects?

Jobs all paid the same?

Knowledge keeps exploding?

Life were a dream?

Light or sound sinks existed?

Limits on immigration to the United States were removed?

Mars was colonized?

Men were twice as big as women?

Mile-high skyscrapers were built?

Money didn't exist?

Money grew on trees?

Mosquitoes could transmit AIDS?

Negative mass existed?

New colors were discovered?

No one was allowed to live beyond a certain age?

Nudity was the norm?

One plus one did not always equal two?

One race was more intelligent than others?

People always lied?

People could change color at will?

People could not see things more than 10 feet away?

People couldn't lie?

People couldn't tolerate others closer than 2 feet?

People didn't feel pain?

People had 360-degree vision?

People had both types of sex organs?

People had elephant trunks?

People had infrared vision?

People had no bones?

People had no hair?

People had skunk glands?

People had two right hands?

People kept getting stupider?

People lost their sense of humor?

People literally attracted and repelled based on feelings?

People spontaneously combusted when angry?

People viewed the world through a fly's eye?

People were cold-blooded?

People were deaf?

Perception was not based on a logarithmic scale?

Plants could communicate?

Plants had feelings?

Pollution got out of control?

Ponds froze from the bottom up?

Population growth continues unchecked?

Prostitution was legal?

Right-handed sugars all became left-handed?

Salaries were all the same?

Satellites could be reached by a staircase?

Sex was a chore?

Silicon-based life existed?

Sleeping caused complete memory loss?

Sound traveled at a speed that varied with frequency?

Sounds left permanent traces on walls?

Space travel becomes routine?

The government had evidence of UFO landings?

The solar system was near the galactic center?

The South had won the Civil War?

The Sun didn't rise tomorrow?

The top of the atmosphere was 100 feet above ground?

The universe was infinite and eternal?

TV disappeared?

Vehicles got a million miles per gallon?

Virginity became fashionable among teenagers?

Voting was compulsory?

Weather could be controlled?

Werewolves existed?

Wheels were square?

Women ruled the world?

World government was established?

You awoke in a different body?

You could double your strength at will?

You could speed up or slow down your own time?

You died?

You felt lightning was about to strike?

You had an elephant for a pet?

You had two weeks to live?

You only saw the world upside down?

You saw a nuclear explosion on the horizon?

You traveled halfway to infinity?

Your dog spoke . . . once?

Bibliography

Recommended for further reading

Dawkins, Richard. *The Blind Watchmaker*. New York: W. W. Norton, 1987.

Ehrlich, Robert. *The Cosmological Milkshake*. New Brunswick, N.J.: Rutgers University Press, 1994.

———. *Turning the World Inside Out, and 174 Other Simple Physics Demonstrations*. Princeton, N.J.: Princeton University Press, 1990.

Epstein, Lewis C., and Paul G. Hewitt. *Thinking Physics*. San Francisco: Insight Press, 1981.

Gamow, George. *Mr Tompkins in Paperback*. New York: Cambridge University Press, 1969.

Hazen, Robert, and James Trefil. *Science Matters*. New York: Doubleday, 1991.

Hewitt, Paul G. *Conceptual Physics*. New York: Harper Collins, 1993.

Hobson, Art. *Physics Concepts and Connections*. Englewood Cliffs, N.J.: Prentice Hall, 1995.

McGrayne, Sharon Bertsch. *365 Surprising Scientific Facts, Breakthroughs, and Discoveries*. New York: John Wiley & Sons, 1994.

Nahin, Paul J. *Time Machines*. New York: American Institute of Physics Press, 1993.

Walker, Jearl. *The Flying Circus of Physics with Answers*. New York: John Wiley & Sons, 1977.

About
the author
and the illustrator

Robert Ehrlich, professor of physics at George Mason University, delights in making physics interesting to students and the general public. He is the author of *The Cosmological Milkshake: A Semi-Serious Look at the Size of Things* (available from Rutgers University Press), *Turning the World Inside Out and 174 Other Simple Physics Demonstrations* (1990), and three other books.

Gary Ehrlich has a degree in architectural engineering and currently works as an acoustical consultant. He also illustrated *The Cosmological Milkshake: A Semi-Serious Look at the Size of Things* and *Turning the World Inside Out and 174 Other Simple Physics Demonstrations*.